D0455035

Love
Sense

ALSO BY DR. SUE JOHNSON

Hold Me Tight: Seven Conversations
for a Lifetime of Love

The Practice of Emotionally Focused Couple Therapy:
Creating Connection

Emotionally Focused Couple Therapy
with Trauma Survivors:
Strengthening Attachment Bonds

Love Sense

. . . .

THE REVOLUTIONARY NEW SCIENCE
OF ROMANTIC RELATIONSHIPS

Dr. Sue Johnson

LITTLE, BROWN AND COMPANY

NEW YORK BOSTON LONDON

Little, Brown and Company
Hachette Book Group
237 Park Avenue, New York, NY 10017
littlebrown.com

First Edition: December 2013

Little, Brown and Company is a division of Hachette Book Group, Inc. The Little, Brown name and logo are trademarks of Hachette Book Group, Inc.

The publisher is not responsible for websites (or their content) that are not owned by the publisher.

The Hachette Speakers Bureau provides a wide range of authors for speaking events. To find out more, go to hachettespeakersbureau.com or call (866) 376-6591.

"If I Should Fall Behind" by Bruce Springsteen. Copyright © 1992 Bruce Springsteen (ASCAP). Reprinted by permission. International copyright secured. All rights reserved.

"i carry your heart with me(i carry it in." Copyright 1952, © 1980, 1991 by the Trustees for the E. E. Cummings Trust, from *Complete Poems: 1904–1962* by E. E. Cummings, edited by George J. Firmage. Used by permission of Liveright Publishing Corporation.

© 1996 "Seasons of Love" (from the Broadway stage play *Rent*) written by Jonathan Larson (1960–1996). All rights reserved. Reprinted courtesy of the Larson Family and Universal Music Publishing, Inc.

Library of Congress Cataloging-in-Publication Data
Johnson, Susan M.
  Love sense : the revolutionary new science of romantic relationships / Sue Johnson.—First Edition.
    pages cm
Includes bibliographical references and index.
ISBN 978-0-316-13376-0 (hardback)
1. Love. 2. Emotions. 3. Interpersonal relations. I. Title.
BF575.L8J636 2013
152.4'1—dc23   2013032372

RRD-C

10 9 8 7 6 5 4 3 2 1

Printed in the United States of America

First, this book is dedicated to my children, in the hope that their love relationships will be richer, deeper, and more fulfilling.

Also, this book is dedicated to those I love—those who continually offer me a safe haven and a secure base from which to joyfully leap into the unknown. You know who you are.

People think love is an emotion. Love is good sense.

—*Ken Kesey*

Unless you love someone, nothing else makes any sense.

—*e. e. cummings*

Contents

Contents

Love Sense

Introduction

We are obsessed with love and love relationships. But what do we really *know* about love? We instinctively know that there is no other experience that will have more impact on our lives—our happiness and health—than our success at loving and being loved. We know that love makes us vulnerable, but also that we are never as safe and strong as when we are sure we are loved. We know that at the most difficult moments in our lives, nothing but the comfort of the ones we love will do. But although, at the end of the first decade of the 21st century, our species has smarts enough to split the atom and soar into space, we still seem to have no clear or rigorous understanding of the intense connection that is so central to our being.

The consensus across human history is that romantic love is, and always will be, an enigma, somehow by its very nature unknowable. I am reminded of Palamon, the imprisoned knight in Chaucer's 14th-century *Canterbury Tales,* who, through his barred window, spies the fair Emily gathering flowers and singing. He cries out in pain and explains to his cousin and fellow prisoner Arcita:

> It's not our prison that caused me to cry.
> But I was wounded lately through the eye
> Down to my heart, and that my bane will be.
> The beauty of the lady that I see
> There in that garden, pacing to and fro,
> Is cause of all my crying and my woe.
> I know not if she's woman or goddess;
> But Venus she is verily, I guess.

Love, to so many of us, seems a bewitchment—a force, powerful and dangerous, that strikes us without our bidding.

Perhaps because love seems so baffling and unruly, we appear to be losing all faith in the viability of stable, romantic partnerships. Pessimism is rampant. On any given day, we scan press accounts and catch videos on TV of famous folks caught in adulterous affairs. We check online advice blogs extolling swinging as the way to combat inevitable relationship fatigue, and read op-ed pieces maintaining that monogamy is an antiquated and impossible concept that should be junked. When it comes to adult love, we appear to be truly lost.

All this at a time when, ironically, romantic love is more important than ever. A tsunami of loneliness, anxiety, and depression is sweeping through Western societies. Today, adult partnerships are often the only real human ties we can count on in our mobile and insanely multitasking world. My grandmother lived in a village of three hundred people that offered her a web of companionship and support, but now most of us seem, if we are lucky, to live in a community of two. Seeking and holding onto a life partner has become a pressing organizing feature of our lives, given that other community ties are so marginalized. The evidence is that we are ever more dependent on our mates for emotional connection and support while being in the dark as to how to create love and keep it.

Moreover, we seem in so many ways to be working actively against our desire for love and commitment. Our society exalts emotional independence. We are constantly exhorted to love ourselves first and foremost. A friend at a social gathering recently commented to me, "Even you have to face it. We are in general a distant and dismissive society. People don't believe in love relationships anymore. They are not the priority. No one has time for them anyway."

As a clinical psychologist, couple therapist, and relationship researcher, I have grown increasingly alarmed and frustrated by where we are and where we seem to be going. Through my own work and that of esteemed colleagues, I know that the cynicism and hopelessness are misplaced. Today, we have a revolutionary new perspective on romantic love, one that is optimistic and practical. Grounded in science, it reveals that love is vital to our existence. And far from being unfathomable, love is exquisitely logical and understandable. What's more, it is adaptive and functional. Even better, it is malleable, repairable, and durable. In short, we now comprehend, finally and irrefutably, that love makes "sense." The word derives from the Latin *sentire,* meaning "to perceive, feel, or know," and also "to find one's way." And that is why I have called this book *Love Sense.* I intend for it to help you find your way to more fulfilling and lasting love.

In *Love Sense,* you will learn what I and other scientists have discovered from thirty years of clinical studies, laboratory experiments, and applied therapies. You will learn that love is a basic survival code, that an essential task of our mammalian brain is to read and respond to others, and that it is being able to depend on others that makes us strong. You will learn that rejection and abandonment are danger cues that plunge us into real physical pain, that sexual infatuation and novelty are overrated, and that even the most distressed couples can repair their bond if they are guided to deal with their emotions a little differently.

My particular contribution lies in relationship repair. Working with thousands of despairing couples through the years has led me to create a new systematic model of treatment, Emotionally Focused Therapy, that honors our need for connection and support. EFT, as it is commonly called, is the most successful approach to healing faltering relationships that has yet been devised, with an astounding 70–75 percent success rate. Today, EFT is routinely taught to counselors in training in at least twenty-five countries around the globe. A simplified version of EFT for couples wishing to help themselves can be found in my previous book *Hold Me Tight: Seven Conversations for a Lifetime of Love.*

These are but a few results of the scientific quest to understand love. In the pages of this book, you will find the results of many more studies as well as the stories of many couples in their most intimate moments. (All stories are composites of several cases and are simplified to reflect general truths. Names and details have been changed to protect privacy.) You will be surprised, and even stunned, at what you read but most of all you will be enlightened, not only as to the nature of love and how it affects us personally but also about what it means to us as human beings and to our society and the world. All the research agrees that a stable, loving relationship is the absolute cornerstone of human happiness and general well-being. A good relationship is better health insurance than a careful diet and a better anti-aging strategy than taking vitamins. A loving relationship also is the key to creating families that teach the skills necessary to maintain a civilized society — trust, empathy, and cooperation. Love is the lifeblood of our species and our world.

Jonathan Larson, the late composer and playwright, put it well in a song from his musical, *Rent,* that asks the measure of "five hundred twenty-five thousand six hundred minutes," or a year in one's existence. The answer: "Share love, give love, spread love... Measure, measure your life in love." Nothing else makes sense.

I write this book not just as a warning but also as a revelation and a promise.

PART ONE

The Relationship
Revolution

CHAPTER 1

Love: A Paradigm Shift

I believe in the compelling power of love. I do not
understand it. I believe it to be the most fragrant
blossom of all this thorny existence.

— Theodore Dreiser

My memories are full of the sounds and sights of love:

The ache in my elderly grandmother's voice when she
spoke of her husband, gone nearly fifty years. A railway
signalman, he had courted her, a ladies' maid, for seven years on the
one Sunday she had off each month. He died of pneumonia on
Christmas Day after eighteen years of marriage, when he was forty-
five and she just forty.

My small enraged mother flying across the kitchen floor at my
father, a former naval engineer in World War II, who stood large
and strong in the doorway, drinking her in with his eyes, and she,
seeing me, stopping suddenly and fleeing from the room. She left
him after three decades of slammed doors and raised fists when I

was ten. "Why do they fight all the time?" I asked my granny. "Because they love each other, sweetie," she said. "And watching them, it's clear that none of us knows what the hell that means." I remember saying to myself, "Well, I won't do this love thing, then." But I did.

Telling my first great love, "I refuse to play this ridiculous game. It's like falling off a cliff." Weeping just months into a marriage, asking myself, "Why do I no longer love this man? I can't even pinpoint what is missing." Another man smiling quietly at me, and I, just as quietly, leaning back and letting myself plunge into the abyss. There was nothing missing.

Sitting, years later, watching the last of the ice finally melting on our lake one morning in early April and hearing my husband and children walking through the woods behind me. They were laughing and talking, and I touched for a moment the deepest joy, the kind of joy that was, and still is, entirely enough to fill up my heart for this lifetime.

Anguish and drama, elation and satisfaction. About what? For what?

Love can begin in a thousand ways—with a glance, a stare, a whisper or smile, a compliment, or an insult. It continues with caresses and kisses, or maybe frowns and fights. It ends with silence and sadness, frustration and rage, tears, and even, sometimes, joy and laughter. It can last just hours or days, or endure through years and beyond death. It is something we look for, or it finds us. It can be our salvation or our ruin. Its presence exalts us, and its loss or absence desolates us.

We hunger for love, yearn for it, are impelled to it, but we haven't truly understood it. We have given it a name, acknowledged its force, cataloged its splendors and sorrows. But still we are confronted with so many puzzles: What does it mean to love, to have a loving relationship? Why do we pursue love? What makes love stop? What makes it persist? Does love make any sense at all?

Down through the ages, love has been a mystery that has eluded everyone—philosophers, moralists, writers, scientists, and lovers alike. The Greeks, for instance, identified four kinds of love, but their definitions, confusingly, overlap. *Eros* was the name given to passionate love, which might or might not involve sexual attraction and desire. In our day, we are equally bewildered. Google reported that the top "What is" search in Canada in 2012 was "What is love?" Said Aaron Brindle, a spokesman for Google, "This tells us about not only the popular topic for that year…but also the human condition." Another website, CanYouDefineLove.com, solicits definitions and experiences from folks around the globe. Scroll through the responses and you'll agree with the site's founders that "there are just as many unique definitions as there are people in the world."

Scientists try to be more specific. For example, psychologist Robert Sternberg of Oklahoma State University describes love as a mixture of three components: intimacy, passion, and commitment. Yes, but that doesn't solve the riddle. Evolutionary biologists, meanwhile, explain love as nature's reproductive strategy. In the grand abstract scheme of existence, this makes sense. But for illuminating the nature of love in our everyday lives, it's useless. The most popular definition is perhaps that love is…a mystery! For those of us—and that is almost all of us—who are trying to find it or mend it or keep it, this definition is a disaster. It robs us of hope.

Does it even matter whether we understand love?

If you had asked that question as recently as thirty or forty years ago, most of the world would have said, "Not really." Love, despite its power, wasn't considered essential to daily life. It was seen as something apart, a diversion, even a luxury, and oftentimes a dangerous one at that (remember Romeo and Juliet and Abelard and Heloise?). What mattered was what was necessary to survive. You tied your life to your family and your community; they provided food, shelter, and protection. Since the earliest conception of marriage, it was understood that when you joined your life to another's, it was for eminently practical reasons, not emotional ones: to better your lot, to acquire power and wealth, to produce heirs to inherit titles and property, to create children to help with the farm and to care for you in your old age.

Even as life eased for growing numbers of people, marriage remained very much a rational bargain. In 1838, well into the Industrial Revolution, naturalist Charles Darwin made lists of the pros and cons of marriage before finally proposing to his cousin Emma Wedgwood. In favor, he noted, "Children...Constant companion, (& friend in old age)...object to be beloved & played with...better than a dog anyhow...a nice soft wife on a sofa with good fire, & books & music...These things good for one's health." Against it, he wrote, "perhaps quarreling—Loss of time.—cannot read in the Evenings...Anxiety & responsibility—less money for books &c...I never should know French,—or see the Continent—or go to America, or go up in a Balloon, or take solitary trip in Wales—poor slave."

We don't have Emma's list, but for most women the top reason to marry was financial security. Lacking access to schooling or jobs, women faced lives of punishing poverty if they remained unwed, a truth that continued well into the 20th century. Even as women gained education and the ability to support themselves, love didn't figure too highly in choosing a mate. When asked in 1939 to rank eighteen characteristics of a future spouse or relationship, women put love fifth. Even in the 1950s, love hadn't made it to first place.

I am reminded of my aunt, who, when she found out that I had a "man in my life," advised me, "Just make sure he has a suit, dear"—code for "Make certain he has a steady job."

In the 1970s, however, love began heading the list in surveys of what American women and men look for in a mate. And by the 1990s, with vast numbers of women in the workforce, marriage in the Western world had completely shifted from an economic enterprise to, as sociologist Anthony Giddens calls it, an "emotional enterprise." In a 2001 U.S. poll, 80 percent of women in their twenties said that having a man who could talk about his feelings was more important than having one who could make a good living. Today, both men and women routinely give love as the main reason to wed. And indeed, this is increasingly the case around the world; whenever people are free of financial and other shackles, they select a spouse for love. For the first time in human history, feelings of affection and emotional connection have become the sole basis on which we choose and commit to a partner. These feelings are now the primary basis for the most crucial building block of any society, the family unit.

A love relationship is now not only the most intimate of adult relationships, it is also often the principal one. And for many it is the only one. The *American Sociological Review* reports that since the mid-1980s, the number of Americans saying that they have *only* their partner to confide in has risen by 50 percent. We live in an era of growing emotional isolation and impersonal relationships. More and more, we dwell far from caring parents, siblings, friends, and the supportive communities we grew up in. And more and more we live alone. According to the latest U.S. census, more than thirty million Americans live solo, compared with just four million in 1950. We toil for longer hours and at more remote locations requiring lengthy commutes. We communicate by e-mailing and texting. We deal with automated voices on the telephone, watch concerts performed by holograms of deceased artists (such as rapper Tupac Shakur), and soon we will be seeking assistance from holographic personnel. At New York City–area airports, travelers were recently introduced to a six-foot-tall, information-spouting AVA, short for airport virtual assistant, or avatar.

Loneliness researcher John Cacioppo, a psychologist at the University of Chicago, contends that in Western societies, "social connection has been demoted from a necessity to an incidental." As a result, our partners have been forced to fill the void. They serve as lover, family, friend, village, and community. And emotional connection is the only glue in this vital, unique relationship.

So yes, understanding the nature of love absolutely does matter. Indeed, it is imperative. Continued ignorance is no longer an option. Defining love as a mystery beyond our grasp and control is as toxic to the human species as is poison in our water. We must learn to shape our love relationships. And now, for the first time, we can, thanks to an unheralded revolution in the social and natural sciences that has been under way for the past twenty years.

A REVOLUTION

Merriam-Webster's Collegiate Dictionary defines *revolution* as "a fundamental change in the way of thinking about or visualizing something: a change of paradigm." And that is exactly what has happened to adult love in the field of social sciences. Two decades ago, love didn't get much respect as a topic of study. No emotion did. René Descartes, the French philosopher, associated feelings with our lower animal nature and thus considered them something to be overcome. What marked us as superior animals was our ability to reason. *Cogito ergo sum*—"I think, therefore I am," he famously proclaimed.

Emotions were not rational and therefore suspect. And love was the most irrational and suspect of all, thus not a fit subject for scientists, the supreme rationalists. Scan the subject index of professor Ernest Hilgard's exhaustive historical review *Psychology in America,* published in 1993; you won't find the word *love.* Young researchers were routinely warned off the topic. I remember being told in graduate school that science does not deal with nebulous, soft indefinables, such as emotion, empathy, and love.

The word *revolution* also means "an uprising." Social scientists began to recognize that much of their work was not addressing public concerns about the quality of everyday life. So a quiet movement, without riots or bullets, began in campus laboratories and academic journals, challenging the orthodox adherence to studies of simple behaviors and how to change them. New voices began to be heard, and suddenly, in the 1990s, emotions emerged as legitimate topics of inquiry. Happiness, sorrow, anger, fear—and love—started appearing on the agenda of academic conferences in a multitude of disciplines, from anthropology to psychology to sociology. Feelings, it was becoming apparent, weren't random and senseless, but logical and "intelligent."

At the same time, therapists and mental health professionals began adjusting their frame of reference in dealing with relationship issues, especially romantic ones. For years they had focused their attention on the individual, believing that any turmoil could be traced back to a person's own troubled psyche. Fix that and the relationship would improve. But that wasn't what was happening. Even when individuals grasped why they acted a certain way and tried to change, their love relationships often continued to sour.

Therapists realized that concentrating on one person didn't give a complete picture. People in love relationships, just as in all relationships, are not distinct entities, acting independently; they are part of a dynamic dyad, within which each person's actions spark and fuel reactions in the other. It was the *couple* and how the individuals "danced" together that needed to be understood and changed, not simply the individual alone. Researchers began videotaping couples recounting everyday hurts and frustrations, arguing over money and sex, and hassling over child-rearing issues. They then pored over these recordings, hunting for the critical moments of interaction when a relationship turned into a battlefield or wasteland. They kept an eye open, too, for moments when couples seemed to reach harmonious accord. And they looked for patterns of behavior.

Interest in emotions in general, and love in particular, also surged among "hard" scientists as advances in technology refined old tools and introduced new ones. A major hurdle to investigations had always been: How do you pin down something as vague and evanescent as a feeling? Or, as Albert Einstein lamented: "How on earth are you ever going to explain in terms of chemistry and physics so important a biological phenomenon as first love?"

The scientific method depends not only on observation and analysis but also on measurable, reproducible data. With the arrival of more sensitive tests and assays, neurobiologists launched inquiries into the chemistry of emotions. But the big push came with the advent of functional magnetic resonance imaging (fMRI). Neurophysiologists devised experiments that peer into the brain and actually *see* structures and areas lighting up when we are afraid, or happy, or sad—or when we love. Remember the old public service announcement showing an egg frying in a pan while a voice intones, "This is your brain on drugs"? Now we have films that actually do capture "This is your brain on love."

The result of all this ferment has been an outpouring of fresh knowledge that is coalescing into a radical and exciting new vision of love. This new love sense is overthrowing long-held beliefs about the purpose and process of romantic love as well as our sense of the very nature of human beings. The new perspective is not only theoretical but also practical and optimistic. It illuminates why we love and reveals how we can make, repair, and keep love.

Among the provocative findings:

• *The first and foremost instinct of humans is neither sex nor aggression. It is to seek contact and comforting connection.*

The man who first offered us this vision of what we now call attachment or bonding was an uptight, aristocratic English psychiatrist, not at all the kind of man you would expect to crack the code of romantic relationships! But John Bowlby, conservative and British, was nevertheless a rebel who changed the landscape of love and loving forever. His insights are the foundation on which the new science of love rests.

Bowlby proposed that we are designed to love a few precious others who will hold and protect us through the squalls and storms of life. It is nature's plan for the survival of the species. Sex may impel us to mate, but it is love that assures our existence. "In uniting the beloved life to ours we can watch over its happiness, bring comfort where hardship was, and over memories of privation and suffering open the sweetest fountains of joy," wrote George Eliot.

This drive to bond is innate, not learned. It likely arose as nature's answer to a critical fact of human physiology: the female birth canal is too narrow to permit passage of big-brained, big-bodied babies that can survive on their own within a short time after birth. Instead, babies enter the world small and helpless and require years of nurturing and guarding before they are self-sustaining. It would be easier to abandon such troublesome newborns than raise them. So what makes an adult stick around and assume the onerous and exhausting task of parenting?

Nature's solution was to wire into our brains and nerves an automatic call-and-response system that keeps child and parent emotionally attached to each other. Babies come with a repertoire of behaviors—gazing, smiling, crying, clinging, reaching—that draw care and closeness from adults. So when a baby boy bawls from hunger and stretches out his arms, his mom picks him up and feeds him. And when Dad coos or makes funny faces at his baby girl, she kicks her legs, waves her arms, and babbles back. And round and round it goes, in a two-way feedback loop.

•*Adult romantic love is an attachment bond, just like the one between mother and child.*

We've long assumed that as we mature, we outgrow the need for the intense closeness, nurturing, and comfort we had with our care-givers as children and that as adults, the romantic attachments we form are essentially sexual in nature. This is a complete distortion of adult love.

Our need to depend on one precious other—to know that when we "call," he or she will be there for us—never dissolves. In fact, it endures, as Bowlby put it, "from cradle to grave." As adults, we simply transfer that need from our primary caregiver to our lover. Romantic love is not the least bit illogical or random. It is the con-tinuation of an ordered and wise recipe for our survival.

But there is a key difference: our lover doesn't have to be there physically. As adults, the need for another's tangible presence is less absolute than is a child's. We can use mental images of our part-ner to call up a sense of connection. Thus if we are upset, we can remind ourselves that our partner loves us and imagine him or her reassuring and comforting us. Israeli prisoners of war report "lis-tening" in their narrow cells to the soothing voices of their wives. The Dalai Lama conjures up images of his mother when he wants to stay calm and centered. I carry my husband's encouraging words with me in my mind when I walk out on a stage to speak.

• *Hot sex doesn't lead to secure love; rather, secure attachment leads to hot sex—and also to love that lasts. Monogamy is not a myth.*

Pick up any men's or women's magazine and you'll find cover lines blaring: SEDUCE HIM! THIS SEXY MOVE WORKS FROM 20 FEET AWAY; 28 THINGS TO TRY IN BED...OR IN A HAMMOCK. OR THE FLOOR; and SEX ACADEMY—GET AN A IN GIVING HER AN O. In our ignorance, we've made physical intimacy the sine qua non of romantic love. As a result, we myopically pour massive amounts of energy and money into spicing up our sex lives. But we have it backward: it is not good sex that leads to satisfying, secure relationships but rather secure love that leads to good—and, in fact, the best—sex. The growing craze for Internet porn is a catastrophe for healthy love relationships precisely because it negates emotional connection.

It is secure attachment, what nature set us up for, that makes love persist. Trust helps us over the rough places that crop up in every relationship. Moreover, our bodies are designed to produce a cascade of chemicals that bond us tightly to our loved ones. Monogamy is not only possible, it is our natural state.

• *Emotional dependency is not immature or pathological; it is our greatest strength.*

Dependency is a dirty word in Western society. Our world has long insisted that healthy adulthood requires being emotionally independent and self-sufficient; that we, in essence, draw an emotional moat around ourselves. We talk of being able to *separate* and *detach* from our parents, our first loved ones, as a sign of emotional strength. And we look with suspicion at romantic partners who display too much togetherness. We say they are too *involved* with, too *close* to, or too *dependent* on one another. In consequence men and women today feel ashamed of their natural need for love, comfort, and reassurance. They see it as weakness.

Again, this is backward. Far from being a sign of frailty, strong emotional connection is a sign of mental health. It is emotional isolation that is the killer. The surest way to destroy people is to deny them loving human contact. Early studies discovered that 31–75 percent of institutionalized children expired before their third birthday. More recent studies of adopted Romanian orphans, many of whom had spent twenty hours a day unattended in their cribs, found that many suffer from brain abnormalities, impaired reasoning ability, and extreme difficulty in relating to others.

Adults are similarly demolished. Prisoners in solitary confinement develop a complex of symptoms, including paranoia, depression, severe anxiety, hallucinations, and memory loss. They call their experience a "living death." "When we isolate a prisoner in solitary confinement," writes Lisa Guenther, associate professor of philosophy at Vanderbilt University and author of *Solitary Confinement: Social Death and Its Afterlives,* "we deprive [him] of the support of others, which is crucial for a coherent experience of the world."

The idea that we can go it alone defies the natural world. We are like other animals—we need ties to others to survive. We see it clearly in a multitude of cross-species combinations: in Thailand, a tiger adopts baby pigs; in China, a dog nurses lion cubs; in Colombia, a cat cares for a squirrel; in Japan, a boar carries a baby monkey on its back; and in Kenya, a giant male tortoise fosters a tsunami-orphaned baby hippo.

We, too, as the Celtic saying goes, "live in the shelter of each other." World War II historians have noted that the unit of survival in concentration camps was the pair, not the individual. Surveys show that married men and women generally live longer than do their single peers.

We need emotional connection to survive. Neuroscience is highlighting what we have perhaps always known in our hearts—loving human connection is more powerful than our basic survival mechanism: fear. We also need connection to thrive. We are actually healthier and happier when we are close and connected. Consistent emotional support lowers blood pressure and bolsters the immune system. It appears to reduce the death rate from cancer as well as the incidence of heart disease and infectious disease. Married patients who have coronary bypass surgery are three times more likely to be alive fifteen years later than their unmarried counterparts. A good relationship, says psychologist Bert Uchino of the University of Utah, is the single best recipe for good health and the most powerful antidote to aging. He notes that twenty years of research with thousands of subjects shows how the quality of our social support predicts general mortality as well as mortality from specific disorders, such as heart disease.

In terms of mental health, close connection is the strongest predictor of happiness, much more so than making masses of money or winning the lottery. It also significantly lessens susceptibility to anxiety and depression and makes us more resilient against stress and trauma. Survivors of 9/11 with secure loving relationships have been found to recover better than those without strong bonds. Eighteen months after the tragedy, they showed fewer signs of post-traumatic stress disorder (PTSD) and less depression. Moreover, their friends considered them more mature and better adjusted than they had been *prior* to the disaster.

• *Being the "best you can be" is really only possible when you are deeply connected to another. Splendid isolation is for planets, not people.*

Like Darwin, with his list of reservations, many of us think of love as limiting, narrowing our options and experiences. But it is exactly the reverse. A secure bond is the launching pad for our going out and exploring the unknown and growing as human beings. It is hard to be open to new experiences when our attention and energy are bound up in worry about our safety. It is much easier when we know that someone has our back. Thus fortified, we become imbued with confidence in ourselves and in our ability to handle new challenges. For example, young professional women who are emotionally close to their partners and seek their reassurance are more confident in their skills and more successful at reaching their career goals. It is an ironic paradox: being dependent makes us more independent.

• *We are not created selfish; we are designed to be empathetic. Our innate tendency is to feel with and for others.*

We are a naturally empathetic species. This part of our nature can be overridden or denied, but we are wired to be caring of others. We are not born callous and competitive, dedicated to our own survival at the expense of others. As biologist Frans de Waal points out, "We would not be here today had our ancestors been socially aloof." We have survived by caring and cooperating. Our brains are wired to read the faces of others and to resonate with what we see there. It is this emotional responsiveness and ability to work together, not our large, thinking brains alone, that has allowed us to become the most dominant animal on the planet. The more securely connected we are to those we love, the more we tune in and respond to the needs of others as if they were our own. Moral decisions and altruistic actions spring naturally from our emotional connection with others.

The bonds of love are our birthright and greatest resource. They are our primary source of strength and joy. Seeking out and giving support are so vital to human beings that social psychologists

Mario Mikulincer and Phil Shaver observe that, rather than being called *Homo sapiens,* or "one who knows," we should be named *Homo auxiliator vel accipio auxilium,* or "one who helps or receives help." To be even more accurate, I say we should be called *Homo vinculum*—"one who bonds."

A UNIFIED THEORY OF LOVE

Understanding that our lovers are our safe haven from the vicissitudes and depredations of life has given us new insights into what makes romantic relationships fail and succeed. For years, all of us have focused solely on what we see and hear. The fights that erupt over money: "You're spending a fortune on shoes you don't need." "All you want to do is save. We're living like misers. There's no fun." The disputes over in-laws: "You're always on the phone with your mother, telling her every little thing we say and do." "You're Daddy's girl, totally. When are you going to grow up?" The disagreements about child rearing: "So he didn't do his homework last night. He gets too much. You're too rigid and controlling." "And you're too lenient. He has no discipline. You let him get away with murder." And the disappointment about sex: "You cheated. How many times? You're such a liar." "Well, I wouldn't have if you were willing to try new things or have sex more often. And anyway, it didn't mean anything."

But concentrating only on what's right before our eyes obscures our vision. We don't get the big picture. Home in on the miniature dots in Georges Seurat's painting and you'll be unaware you're seeing *A Sunday on La Grande Jatte.* Sit at the piano and play a few notes in a score and you won't hear Johannes Brahms's lulling Waltz in A-flat Major. Take the dance floor and repeat one series of steps and you'll never realize the sensuality of Argentine tango.

Similarly, troubled couples are fixated on specific incidents, but the true problem is broader and deeper. Distressed partners no longer see each other as their emotional safe haven. Our lover is supposed to be the one person we can count on who will always respond. Instead, unhappy partners feel emotionally deprived, rejected, even abandoned. In that light, couples' conflicts assume their true meaning: they are frightened protests against eroding connection and a demand for emotional reengagement.

In contrast, at the core of happy relationships is a deep trust that partners matter to each other and will reliably respond when needed. Secure love is an open channel for reciprocal emotional signaling. Love is a constant process of tuning in, connecting, missing and misreading cues, disconnecting, repairing, and finding deeper connection. It is a dance of meeting and parting and finding each other again, minute by minute and day by day.

The new science has given us what I like to call a unified field theory of love. Einstein couldn't find it for physics, but we've found it for love. At last, all the pieces we've been puzzling over separately fit together. We see the grand scheme. Fifty years ago noted animal researcher Harry Harlow, in an address to the American Psychological Association, observed, "As far as love or affection is concerned, psychologists have failed in their mission...The little we write about it has been better written by poets and novelists."

Today we have cracked the code of love. We now know what a good love relationship looks and feels like. Even better, we can shape it. For the first time, we have a map that can guide us in creating, healing, and sustaining love. This is a consummate breakthrough. At last, to quote Benjamin Franklin, this "changeable, transient, and accidental" phenomenon—romantic love—can be made more predictable, stable, and deliberate.

The fixes we've tried in the past have been failures because we have not understood the basis of love. In general, therapists have attacked the problem in two ways. The first is analytical: couples dig back and sift through their childhood experiences to find the reasons why they respond the way they do. This seeking after insight into first relationships is laborious, time consuming, and expensive—with small benefit. It comes at the problem sideways, through intellectual insight into each person's relationship history. Your present relationship is not just your past automatically playing out; this dismisses your partner and the power of his or her responses, as if this partner were simply a blank screen on which you project the movie of your past.

The second approach is practical. Couples are instructed on how to communicate more effectively—"Listen and repeat back what your partner has said." Or they're taught how to negotiate and bargain their way through divisive issues, from sex to cleaning—"You agree to vacuum the rug, and I'll clean the bathroom." Or coached on how to improve their sex life—bring on the flowers and racy lingerie and try positions from the Kama Sutra. All of these techniques can be helpful, but only temporarily. Love is not about whether you can parrot back what's said or decide who vacuums the rug or agree on what sexual moves to try. Such practical counseling is like putting a finger in a cracked dam to hold back the tide or sticking a Band-Aid on a suppurating wound.

My client Elizabeth tells me, "The other therapist made us do these set exercises using the statements she gave us, but we just couldn't talk to each other that way when we got home, let alone when we were upset. And we did make a deal about chores, but it didn't change the way I felt about us. I was still lonely. At one point we were doing this 'leave the room, take time out' thing, but then I was even more angry when he walked back in, and I didn't even really know what I was so angry about."

Ultimately, these remedies are ineffectual because they don't address the source of relationship distress: the fear that emotional connection—the font of all comfort and respite—is vanishing.

When we know how something works, fixing it and keeping it healthy is much easier. Before this basic understanding, all we could do was flail around trying to fix one part of the relationship in the hope that trust and loving connection would somehow find their way back in through these narrow routes. The new science has given us a straight arterial road to our destination.

To really help couples find happiness, we must shore up the foundation of their relationship; that is, help them relay and rebuild their emotional connection. The technique I and my colleagues have devised, EFT, or Emotionally Focused Therapy (my irreverent children call it Extremely Funny Therapy), does just that. We've discovered that discontented lovers fall into set patterns of behavior that plunge them into cycles of recrimination and withdrawal. The key to restoring connection is, first, interrupting and dismantling these destructive sequences and then actively constructing a more emotionally open and receptive way of interacting, one in which partners feel safe confiding their hidden fears and longings.

The results of EFT, as measured in a multitude of studies, have been astoundingly positive—better, in fact, than the outcomes of any other therapy that has been offered. Lovers say that they feel more secure and satisfied with their relationship. Their mental health improves as well; they are less depressed and anxious. And they are able to hold onto the changes they make long after therapy has ended.

Why is EFT so effective? Because it goes to the heart of the matter. We do not have to persuade or coach partners to be different. The new science has plugged us into the deepest human emotions and opened the way to transfiguring relationships, using the megawatt power of the wired-in longing for contact and care that defines our species. Says one of my clients: "For twenty-eight years, my wife and I have been circling around the kind of conversation we are having now, but we've never actually gotten down to it...Either we were too afraid or we didn't know how. This conversation changes everything between us."

Once you have a map to the territory called love, you can put your feet on the right path and find your way home.

꩜

To help you turn the new science into love sense, you'll find brief "experiments" for you to do at the end of each chapter. Science, after all, is deliberate observation that leads to identification of recurring patterns. By doing these experiments, you'll be collecting data on your own relationship that will help you understand the way you love and help you find the security and satisfaction you—and we all—long for.

EXPERIMENT

Find a quiet place where you will not be interrupted for about thirty minutes. Sit comfortably and quietly, and count twenty breaths in and out. Now imagine that you are in an unfamiliar, dark place. You are suddenly unsure and scared and aware that you are very much alone. You want to call out for someone to come.

Step 1
Who is the person you want to respond to you? Imagine his or her face in your mind's eye.

Do you call or not? Perhaps you convince yourself that this is a bad idea, even a sign of weakness, or an opening that will lead to hurt and disappointment. Perhaps you decide that it is not good to rely on another person and that you must take care of your distress on your own, so you hunker down in the dark. Perhaps you call, but very hesitantly, then go hide in a dark corner.

If you call, how do you do it? What does your voice sound like? When someone comes, what does he do? Does he express concern, offer comfort and reassurance, and stay with you so that you relax and let yourself be comforted?

Or does she come, but then sometimes turn away, dismiss your distress, tell you to control your emotions, or even criticize you, so that you try to hold onto her but get more upset, feeling that she has not really heard your call or cannot be relied upon?

How does your body feel as you do this experiment? Tight, numb, sore, agitated, calm, relaxed? How hard was it for you to do this experiment? Do any emotions come up for you—sadness, joy, anger, or even anxiety?

Step 2

Now stand up and move around for a few minutes. Sit in another chair to consider the results of your thought experiment from some distance. (If it is hard to get distance, you may want to postpone reflecting on the experiment until another day or even discuss it with someone you trust.)

Summarize, in very simple terms, what happened in this fantasy scenario. Write the steps down. What does this imagined scenario tell you about what you expect in a relationship? Our expectations, our predictions about how others will respond to us guide our steps in any dance with a lover. They are our very own love story.

Step 3

Reflecting a little more, see if you can articulate your general feeling about love relationships.

Some people automatically go to phrases such as: "They just don't work"; "Men/Women are impossible to relate to. They always reject you or let you down"; "Love is hard work, but it's worth it"; or "Love is for dummies."

Step 4

Ask yourself, "What do I really want to know about love and loving?" See if you can find the answer by reading the rest of this book.

CHAPTER 2

Attachment: The Key to Love

Love consists in this, that two solitudes protect and
touch and greet each other.

—*Rainer Maria Rilke*

L ove affairs are just rational bargains," lectured a famed psychologist thirty years ago at an international conference in
Banff. "They're negotiations about profit and cost. We all
want to maximize our profit." Sitting in the audience as a newly
minted clinician-researcher, I shook my head. I had been working
with distressed couples, and I knew they didn't fit into this fash-
ionable "exchange theory" of love. But I didn't know why. Hours
later I was sitting in a bar, arguing with a senior colleague. "What's
wrong with the idea? Love relationships *are* rational bargains," he
insisted. "No, they're not," I maintained. "Okay; if they're not,
what are they?" he shot back. I was blank for a moment, then
blurted out excitedly, "They're not bargains. They're bonds. Emo-

tional bonds. Just like the ones between mother and child. Just like John Bowlby said."

CHILDREN AND LOVE

Every revolution has its heroes, and in the relationship revolution, John Bowlby is the hero. Chances are, until Chapter 1, you'd never heard the name Bowlby, but his vision and work have already radically reshaped our relationships with our children and are now doing the same in our relationships with our romantic partners. Bowlby, a British psychiatrist, is the father of attachment theory, a developmental perspective on personality that puts our emotions and our interactions with loved ones front and center in terms of who we are and how we behave.

Over the past forty years, the attachment perspective has seeped into our culture and changed the way we rear our children. It is not so long ago that child-care experts were advocating distant, detached care, the point of which was to turn children into self-possessed, autonomous beings as quickly as possible. One of the fathers of modern behaviorism, John B. Watson, was adamant that mother love was a "dangerous instrument"; women's sentimental natures were a defect that prevented them from pushing their children into independence. Showing warmth, by hugging and cuddling, for example, warped children and made them into weak, emotionally labile adults. If, on the other hand, children were left to cry themselves to sleep, they learned to control themselves and tolerate discomfort. Watson was about as wrong as he could be, although his basic idea—that responding to people's emotional needs makes them more needy, immature, and hard to love—is still very popular when applied to adults.

The majority of us now explicitly recognize a child's need for ongoing, reassuring physical and emotional connection with his or her parents. We acknowledge the power of parental responsiveness in shaping a child's personality. There are some who still argue that loving care is fine, but the roots of personality are indelibly set by our genetic heritage. But this is not so. Study after study has shown that even when genetic heritage is totally stacked in a negative direction, it is our primary relationships that decide if genes come online and how they play out. Highly agitated monkeys, the future bad boys of their tribe, if cared for by especially nurturing foster moms, turn into respected leaders.

Add to genetic problems a stressful environment, and still the responsiveness of the parent makes a difference. Very irritable infants born into poverty often have difficulty controlling their moods, calming themselves, and signaling their needs to their mothers. Researchers at the University of Amsterdam gave mothers of such infants six hours of instruction in recognizing babies' signals and prolonging soothing activities, such as holding and stroking. The improvement was startling. By twelve months of age, the infants matched normal babies in their ability to turn to their mothers for comfort when they were upset and to calm down when soothed by them. In another group, one in which the mothers were not counseled, only 28 percent of the children were rated as securely attached. Connection and care matter.

The revolution in child care came first from simple observation of responses and patterns of interaction between mother and child and then from experiments that set up and manipulated these patterns. (We will see later that the explosion of discoveries in adult bonding initially happened in the same way.) In the 1930s and 1940s, doctors noted that large numbers of orphaned children, who were fed and sheltered but deprived of touch and emotional support, were dying, often before the age of three. Psychoanalyst René Spitz coined the term "failure to thrive" to describe these children. Other health-care workers, meanwhile, were identifying youngsters who were physically healthy but alienated and unable to connect with others. Psychiatrist David Levy suggested they suffered from "emotional starvation."

But it took John Bowlby to really grasp the enormous import of these facts. Born in 1907 to a British baronet and his wife, Bowlby, the fourth of six children, was reared in typical upper-class fashion. He and his siblings saw their parents sparingly. Scrubbed and dressed, they joined their mother for one hour each afternoon for tea; they saw their father, a surgeon, once a week on Sunday. The rest of their time was spent mainly with nursemaids, nannies, and governesses. Bowlby was especially fond of one nursemaid, Minnie, who had been his main caretaker. His mother dismissed her when he was four, a split he later described as being as painful as losing a mother. At the age of seven, he was sent to boarding school, an event so traumatic that years later he told his wife, Ursula, that he wouldn't send a dog to English boarding school at that age.

These experiences seem to have sensitized Bowlby to children's relationships with parents and other significant adults. After Trinity College, Cambridge, where he studied psychology, Bowlby worked at progressive residential schools with maladjusted and delinquent youngsters, many of whom early on had been neglected or separated from their parents. Bowlby went on to become a physician and then a psychoanalyst. He soon found himself in conflict with analytic orthodoxy, which, following Freud's teachings, held that patients' problems were almost invariably internal, traceable to their own unconscious fantasies and struggles. From his own experiences and reports by others, Bowlby was convinced that many patients' difficulties were the opposite, in fact caused by their real relationships with other people. In 1938, as a novice clinician, he was assigned the case of a hyperactive three-year-old boy. His supervisor was the acclaimed analyst Melanie Klein. Bowlby wanted to explore the child's relationship with his extremely anxious mother, but Klein considered only the boy's fantasies about his mother important and forbade him to even speak with the woman. Bowlby was outraged.

Continuing to work with disturbed youngsters, Bowlby came to believe that disrupted relationships with parents or surrogate care-givers could cripple healthy emotional and social growth, producing alienated and angry individuals. In 1944, Bowlby published a seminal article, "Forty-Four Juvenile Thieves," observing that "behind the mask of indifference is bottomless misery and behind apparent callousness, despair." He expanded upon his findings in a groundbreaking study of European children who were evacuated from their homes or left orphaned by World War II. Undertaken at the request of the World Health Organization and published in 1951, the study concluded that separation from loved ones deprived youngsters of emotional sustenance and was as damaging to the psyche as lack of food is to the body.

The work brought both condemnation and praise. Bowlby focused on the mother-child bond; he carefully noted that "the infant and young child should experience a warm, intimate, and continuous relationship with his mother (or permanent mother substitute) in which both find satisfaction and enjoyment." Feminists complained that Bowlby's claim chained women to constant child tending and denied their need to go out into the world and have independent lives. Government officials, meanwhile, applauded. Many returning veterans were unemployed, their jobs having been filled by women during the war effort. Here was a reason to move women out and a way to move men back in.

Bowlby's developing theory was controversial in another way. It marked a further break with accepted dictum. Freud had maintained that the link between mother and child is forged after birth and is a conditioned response. Baby loves Mom because she rewards him with food. But Bowlby, who was impressed by Darwin's theory of natural selection and the work of contemporary ethologists, was convinced that the emotional tie is wired in before birth and automatic. Support for Bowlby's thesis came from the dramatic experiments of his colleague and friend Harry Harlow, a psychologist at the University of Wisconsin, who was studying rhesus monkeys separated from their mothers at birth. Raised in isolation, the baby monkeys were so hungry for "contact comfort" that when presented with a choice between a wire "mother" who dispensed food and one made of soft rags that didn't dispense any fare, they almost always clung to the spongy impersonator. As science writer Deborah Blum observes in her book on the work of Harry Harlow, food is sustenance, but a good hug is "life itself."

In an attempt to prove his ideas, Bowlby collaborated with James Robertson, a young social worker, to make a documentary called *A Two-Year-Old Goes to Hospital.* The film tells the story of a young girl named Laura, who goes to the hospital for a minor operation and stays for eight days. The film is horrifying. (You can view portions on the Internet, and I guarantee you will be in tears.) Following the era's prevailing professional wisdom—that coddling by mothers and other family members creates clinging, dependent children who grow into ineffectual adults—parents were not permitted to stay with their hospitalized offspring. Sick sons and daughters had to be dropped off at the door; parents were allowed to visit for one hour a week.

Separated from her mother and faced with a revolving cast of nurses and doctors, Laura is frightened, angry, hysterical, and, finally, totally desolate. When she is at last released from the hospital, she is emotionally shut down and completely withdrawn from her mother. The film caused a sensation in professional circles. The Royal Society of Medicine denounced it as a fraud, and the British Psychoanalytical Society dismissed it, with one analyst declaring that Laura's grief and terror was induced not by separation from her mother but by unconscious angry fantasies concerning her mother's new pregnancy. (It wasn't until the late 1960s that British and American hospitals abandoned their rigid policies and allowed parents to stay with their children.)

Despite rejection by the establishment, Bowlby pioneered on, giving form to a theory of what he called *attachment*. (The story goes that when asked by his wife why he didn't give it its rightful name, a theory of love, he replied, "What? I'd be laughed out of science.") Bowlby was aided significantly in his work by psychologist Mary Ainsworth, a Canadian researcher who helped give shape to his ideas and test them.

Together, they identified four elements of attachment:

- We seek out, monitor, and try to maintain emotional and physical connection with our loved ones. Throughout life, we rely on them to be emotionally accessible, responsive, and engaged with us.
- We reach out for our loved ones particularly when we are uncertain, threatened, anxious, or upset. Contact with them gives us a sense of having a safe haven, where we will find comfort and emotional support; this sense of safety teaches us how to regulate our own emotions and how to connect with and trust others.
- We miss our loved ones and become extremely upset when they are physically or emotionally remote; this separation anxiety can become intense and incapacitating. Isolation is inherently traumatizing for human beings.
- We depend on our loved ones to support us emotionally and be a secure base as we venture into the world and learn and explore. The more we sense that we are effectively connected, the more autonomous and separate we can be.

The above four elements are considered to be the norm and universal, occurring in relationships across cultures. The basic concept is that forming a deep mutual bond with another is the first imperative of the human species. As Bowlby saw it, life at its best is essentially a series of excursions from the safety of a secure relationship out into the uncertainty of the greater world.

Bowlby's theory was missing empirical evidence, however. Mary Ainsworth came to the rescue. She devised a simple and ingenious experiment that is regarded as one of the most important and influential in all of psychology. It is as crucial to our understanding of love and relationships as Newton's experiment showing that pebbles and heavy rocks fall at the same speed is to our understanding of gravity and the physical world. In truth, if not for Ainsworth's experiment, Bowlby's idea might still be just supposition.

The experiment is called the Strange Situation, and you can see variations of it on the Internet. A mother and her toddler are in an unfamiliar room. A few minutes later, a researcher enters and the mother exits, leaving the youngster alone or with the researcher. Three minutes later, the mother comes back. Most children are initially upset at their mother's departure; they cry, throw toys, or rock back and forth. But three distinct patterns of behavior emerge when mother and child are reunited—and these patterns are dictated by the type of emotional connection that has developed between the two.

Children who are resilient, calm themselves quickly, easily reconnect with their moms, and resume exploratory play usually have warm and responsive mothers. Youngsters who stay upset and nervous and turn hostile, demanding, and clingy when their moms return tend to have mothers who are emotionally inconsistent, blowing sometimes hot, sometimes cold. A third group of children, who evince no pleasure, distress, or anger and remain distant and detached from their mothers, are apt to have moms who are cold and dismissive. Bowlby and Ainsworth labeled the children's strategies for dealing with emotions in relationships, or attachment styles, secure, anxious, and avoidant, respectively.

Bowlby lived to see his attachment theory become the cornerstone of child rearing in the Western world. (Indeed, the term *attachment parenting* has become so accepted and widespread that it has been affixed to an intense form of parenting recommended by pediatrician William Sears. Though it is based on Bowlby's tenets, it goes far beyond anything he advocated. In attachment parenting, children often sleep in the parental bed, breastfeed for several years, and are, generally, in almost constant contact with their mom or dad.)

Today, no one doubts that youngsters have an absolute need for close emotional and physical contact with loved ones. That perspective has become part of the air we breathe, but only when we think of childhood. Many of us still believe that adolescence ends such dependence. Bowlby did not. He maintained that the need to be close to a few precious others, to attach, persists through life and is the force that shapes our adult love relationships. As he wrote: "All of us, from cradle to grave, are happiest when life is organized as a series of excursions, long or short, from the secure base provided by our attachment figure(s)."

ADULTS AND LOVE

Bowlby based his claims in part on his observation of World War II widows, who, he found, showed the same behavior patterns as homeless orphans. He was also well aware that the isolated monkeys in Harlow's experiments who lived to maturity were emotional wrecks, lapsing into self-mutilation, rage, or apathy and failing to relate to other monkeys and to mate. But again and again, his ideas were rebuffed. Bowlby died in 1991, before he could assemble evidence that his attachment perspective was indeed relevant to adults and adult love relationships.

Phil Shaver and Cindy Hazan, social psychologists then at the University of Denver, took up the torch a few years later. They were initially interested in how people handled grief and loneliness, and they began to read Bowlby's research, looking for clues about why loneliness was so devastating. Bowlby's work so impressed them that they decided to put together a quiz about love and relationships that appeared in the *Rocky Mountain News*. The survey, although unscientific, indicated that the same attachment characteristics and behavior patterns that occur between mothers and children also occur between adults. When lovers felt secure, they could reach out and connect, easily helping each other find their emotional balance; but when they felt insecure, they became either anxious, angry, and demanding or withdrawn and distant. Shaver and Hazan launched more formal studies, and their work inspired others to test Bowlby's predictions.

In the two decades since Bowlby's death, hundreds of studies have been published bearing out his assertions. They confirm that our need to attach continues beyond childhood and also establish that romantic love is an attachment bond. At every age, human beings habitually seek and maintain physical and emotional closeness with at least one particular irreplaceable other. We especially seek out this person when we feel stressed, unsure, or anxious. We are just hardwired this way.

The fact that this perspective on adult love was, at first, summarily rejected by many psychologists and mental health professionals is not surprising. For one thing, it challenges a cherished belief about ourselves as adults; specifically, that we are self-sufficient entities. (We are bombarded daily with media messages — "Love yourself!" "You're worth it!" — and instructions on how to soothe ourselves.) Bowlby's belief also goes against an increasingly popular conception of love relationships: that they are essentially companionships with sex added. But those of us who flourish, even when living alone, invariably have a rich internal world populated by images of loving attachment figures. To be human is to need others, and this is no flaw or weakness. And being friends, even with physical intimacy, is different from being lovers. The bond with a friend is not as tight. No matter how close, friends cannot offer the degree of caring, commitment, trust, and safety that true lovers do. They are not our irreplaceable others.

In a recent experiment, psychologist Mario Mikulincer of the Interdisciplinary Center (IDC) in Herzliya, Israel, asked students to name people they love and people they are simply acquainted with. Then he gave them a task on a computer. They had to look at a list of letters, tap one key if a string of letters could be formed into a word or a name, and tap another key if the letters could not. At times during the task, threatening words, such as *separation, death,* and *failure,* flashed on screen, but only for milliseconds (much too fast for conscious processing). What Mikulincer found was that after the subliminal threats, students sorted the scrambled letters of the names of their loved ones much faster than the names of acquaintances and friends.

In psychology, reaction time in recognizing words is a commonly used measure of the accessibility of a person's thoughts. The quicker the reaction time, the higher the accessibility. This study shows that if we're primed with any kind of threat, we automatically and swiftly pull up the names of our loved ones—they are our safe haven. This experiment reminds me of everyday situations. My husband's first thought when he got the date for a medical test was whether I would be home that day and free to accompany him. I know that when I'm on a plane landing in a storm, I automatically call up the image of the slow smile my husband will give me at the gate. In the next chapter, you'll hear about an fMRI study in my lab in which women who expected to get a small electric shock experienced much less fear simply by holding their husband's hand.

Bowlby and Ainsworth said children bond to their loved ones in three ways, and this is true of adults as well. A person's basic attachment style is formed in childhood. Secure, the optimal style, develops naturally when we grow up knowing that we can count on our main caregiver to be accessible and responsive to us. We learn to reach for closeness when we need it, trusting that we will be offered comfort and caring much of the time. This loving contact is a touchstone, helping us to calm ourselves and find our emotional balance. We feel comfortable with closeness and needing others and aren't consumed by worry that we will be betrayed or abandoned. Our behavior says, in essence, "I know I need you and you need me. And that's okay. In fact, it's great. So let's reach out to each other and get close."

Some of us, however, had early caregivers who were unpredictably or inconsistently responsive, neglectful, or even abusive. As a result, we tend to develop one of two so-called insecure strategies—anxious or avoidant—that automatically turn on when we (or our partners) need connection. If we have an anxious style, our emotions are ramped up; we are inclined to worry that we will be abandoned, and so we habitually seek closeness and ask for proof that we are loved. It's as if we are saying, "Are you there? Are you? Show me. I can't be sure. Show me again."

If we have an avoidant style, on the other hand, we tend to tamp down our emotions so as to protect ourselves from being vulnerable to, or dependent on, others. We shut down our attachment longings and try to evade real connection. We are apt to see other people as a source of danger, not safety or comfort. Our attitude seems to be "I don't need you to be there for me. I'm fine whatever you do."

Although we have a main attachment style, we can—and do—

step into alternative strategies at specific times and with specific people. In my own interactions with my husband of twenty-five years, I am secure most of the time, but if we have been at odds for a while, I can slip into a more anxious style, demanding that he respond and soothe my disquiet. When he does, then I go back to my primary, secure strategy.

Just for fun, I've picked three of my English relatives to illustrate the three basic styles. My father, Arthur, had a secure style. He listened when I, his only child, announced I was going to Canada, told me how much he would miss me, and then asked me what I needed. He gave me the encouragement I was longing for, and also told me that I could always come back home to him if things didn't work out. He also wrote me regular, loving letters. He freely offered support to others as well. A naval engineer on destroyers in World War II, he opened his arms, literally, to other veterans, holding them in the back room of our family's pub while they cried over lost friends and devastated lives. My father knew, too, how to seek support for himself. He asked his best buddy to accompany him to the hospital when he went in for an operation on his back.

My lanky Auntie Chloe, who looked exactly like Popeye's love, Olive Oyl, had a highly anxious style. She thought my small, portly Uncle Cyril, with his Elvis Presley pompadour, was fatally attractive to other women and that even his potbelly added to his sexual allure. He went away on business often, and when she talked about this, Auntie Chloe would tear up and openly wonder if he were having what she called "lascivious liaisons." His habitual silence when he was home did nothing to reassure her. She would hang on to his arm at family gatherings as if he were about to vaporize. Even back then, I thought that she might have been less clingy if he had been a little more open and talkative. After all, he was hard to know, and I never felt any real sense of safety with him, either.

Tall, gruff Uncle Harold was extremely avoidant. When I went to stay at his home and burst into tears because my teddy bear had become filthy from the mud pies I fed him and then had come apart when I scrubbed him with toilet cleaner, Harold told me, "Cut that soppy stuff," and sent me to my room. He was unapproachable, especially by little girls, and usually spent his days in the garden and often slept on the pull-out bed in the shed. When I was present, he never touched Vina, his friendly, jolly wife of thirty years. Still, he nursed her when she became ill, and three months after she died, he committed suicide. "He couldn't be close, but he just couldn't live without her," my granny told me.

Attachment styles line up neatly with the basic way we see ourselves and others. These "mental models" shape the way we regulate our emotions, and they guide our expectations in love relationships, assigning meaning to our partner's actions and becoming "If this, then that" templates for how to interact. Secure people see themselves as generally competent and worthy of love, and they see others as trustworthy and reliable. They tend to view their relationships as workable and are open to learning about love and loving. In contrast, anxious people tend to idealize others but have strong doubts as to their own value and their basic acceptability as partners. As a result, they obsessively seek approval and the reassurance that they are indeed lovable and not about to be rejected. Avoidant folks, meanwhile, view themselves as worthy of love—at least that is their conscious stance. Any self-doubt tends to be suppressed. They have a negative view of others as inherently unreliable and untrustworthy. Even in their stories and dreams, anxious people portray themselves as apprehensive and unloved, while avoidants see themselves as distant and unfeeling.

Psychologist Jeff Simpson is doing watershed studies in this area. Jeff, who looks like an all-American, crew-cut quarterback, speaks with me from his lab at the University of Minnesota. As a kid he loved watching people interact, especially at the medical clinics, where he went regularly to get his allergy shots. He remembers being fascinated by the fact that, when they looked scared or sad, some folks wanted to talk, some wanted to be touched and hugged, and some wanted to be left alone. He recalls, too, as a college kid, being on special assignment in Oxford, England to study the behavior of farm cats. He got hooked on finding patterns in interactions. So it seemed natural that he decided to become a social psychologist.

But once in graduate school, in the early 1990s, he was disappointed. He discovered that most psychologists weren't studying face-to-face interactions; rather, they were asking adults to fill in questionnaires that collated opinions and attitudes that rumbled around in their brains. A very few researchers were trying to look at how people under stress actually behave in relation to others, but they couldn't explain what they were seeing. Jeff had hit a dead end. Steve Rholes, a fellow graduate student, came to the rescue. He pointed out that John Bowlby had a theory, supported by studies with babies, that might also apply to adults and explain why some reach out for support when they are upset and others turn away. Over coffee, they decided to set up an experiment to see what people in dating relationships would do when placed in an upsetting situation. Ta-da! The first observational study of attachment behavior in adults was born.

Jeff and Steve asked heterosexual couples to fill out questionnaires and rate statements such as "I find it relatively easy to get close to others" in order to assess the partners' attachment styles. The researchers then told the couples that the female partner would soon be placed in a nearby room to engage in an unspecified activity that creates anxiety in most people. They showed the couples the room, dark and full of ominous-looking equipment, and left them waiting outside. A video camera secretly recorded the couples over the next five minutes. Researchers analyzed the tapes, looking for support-seeking and support-giving behavior.

"I knew we were going to get really interesting results when we watched the video of one of the first couples," Jeff tells me. "This woman had been jovially chatting with her partner before she was told about the 'activity,' but after, when her partner, looking concerned, asked if she was okay and reached out to her, she said, 'Leave me alone,' and moved away. Then he said, 'Can I help?' and she exploded. She turned and hit out at him, pushed her chair away, and grabbed a magazine. When we went back and looked at her attachment style test, she scored as extremely avoidant. We had found a way to link a person's attachment style, arising from their history with others, to their present expectations and their way of dealing with their emotions. And all this predicted specifically how they behaved in their love relationships when faced with a stressful situation. It was obvious to us that these kinds of links played a big part in defining the nature of specific love relationships."

Other studies by Jeff's team have also confirmed that, just as Bowlby predicted, secure and anxiously attached people tend to reach for those they love for comfort while avoidant people tend to withdraw. But then they discovered a wrinkle. That finding is true only when the threat comes from *outside* the relationship, as in the study above. When it comes from *inside,* the responses are different. Both secure and avoidant people can stay on topic and keep their emotions in check while discussing internal conflicts—say, the fact that one partner wants more sex than the other—although secure folks are still better at constructing solutions and acting warmly toward their partner. But in the face of internal conflict, anxious partners do not reach out; they go completely off the rails. They catastrophize, bring in irrelevant issues, and become angry and confrontational, even when their partner refrains from being reciprocally hostile. Anxious partners are generally uneasy about their lover's commitment to begin with and thus are primed to view anything he or she says or does more negatively. Haunted by the specter of abandonment, they try to control their lover.

Such face-to-face studies marked a huge shift in our understanding of love relationships. Before that, a lot of our "knowledge" came through stories, tales, and poems, or the age-old way, through gossip and platitudes (these are still popular sources of "wisdom" about love, only now they are carried on the Internet). Jeff tells me, "I wanted to show people that psychology now can help us understand not just what is between our ears but also what is between one person and another. We can study Jack and Jill as they interact and lay out the structure of adult human bonds."

The way we attach as adult lovers tends to reflect the way we attached as children to our primary caregivers. Jeff's group has looked at data collected in the Minnesota Longitudinal Study of Risk and Adaptation. The project, begun in the 1970s and led by psychologist Alan Sroufe, has been following more than 200 people as they have developed from birth into adulthood. Jeff's team has discovered a consistent thread running through people's relationships with their mothers, their first adult romantic partner, and their later lovers. The more securely attached to their mother the subjects were as children, the more secure were their attachments to others at later stages of development. And, notes Jeff, the strength of their tie to their mom at age one predicted how good they were at dealing with their emotions and resolving conflicts with their adult partners at age twenty-one.

Love is also, inevitably, about loss. Deborah Davis at the University of Nevada, Reno, has spearheaded research that demonstrates that attachment patterns have an impact even when couples are splitting up. Through the Internet, she asked 5,000 people to respond to questions about their attachment style and their behavior during breakups. Davis's study grew out of her personal experience; her marriage was breaking up, and her spouse was seesawing between expressions of adoration and rage. She started thinking about Bowlby's descriptions of attachment distress in children—angry protest, clinging and seeking, and depression and despair—and decided to look at whether adults' behavior during a romantic turmoil could be predicted.

She expected more anxiously attached people to be more frantic and to try to pull their partner back by making demands and threats. And that's what she found. Compared to basically secure individuals, anxiously attached partners described becoming more obsessed and angry and committing more hostile, threatening acts, like destroying property. They also described feeling more longing and having more sexual desire for the partner who was leaving. This fits with the attachment perspective—that anxious partners often show "rejection sensitivity," both expecting dismissal and reacting to it with increased aggression. Other researchers have found that perceived rejection triggers violence, especially in more anxiously attached male partners.

Avoidants cope by doing things to lessen contact with the rejecting partner, such as moving out of the area where they were together. They hunker down and turn inward, relying on themselves in these situations. They don't talk to friends but try to distract themselves, pulling away from reminders of the relationship and suppressing their distress. More avoidant folks also tend to steer clear of new relationships for a while, whereas some anxious partners try to jump into new relationships immediately. Both highly anxious and avoidant people do have one similarity: they often resort to alcohol and drugs as a way of coping with romantic turmoil, more so than do people who are basically secure.

When I consider all this, I find myself recalling an old lover breaking into my flat many years ago and leaving nasty messages everywhere about how terrible I was and how I would forever regret sending him away. Months later I would open a book, and a barbed missive from him would float to the floor to wound me yet again. There are only so many ways to deal with the helplessness and hopelessness we feel when we lose the person we have bonded to.

The irony here is that when we are able to have a more secure bond with a partner, we not only love better, we deal with the loss of love better. My clients tell me sometimes that they are afraid to love because of the risk of devastation if a loved one leaves, but in fact secure connection is linked to faster emotional recovery from the loss of a partner as well as to less sadness and anger. When we attain a secure loving bond with another, we can, in a sense, keep that felt sense of connection with us even after that person has, for whatever reason, exited from our lives. My more secure friends seem to talk about old lovers without rancor and with a positive sense of what they gave and received in those relationships.

In sum, we can see attachment theory and science as offering us an architecture of romantic love. Think of yourself as a house. On the first floor and reaching into the foundation are your basic needs for comfort, reassurance, connection, closeness, and care as well as your basic emotions, including joy, fear, sadness, and anger. These are wired in by thousands of years of evolution. On the second floor are your ways of coping with these needs and emotions, opening to and trusting them, cutting them off or defending against them, or becoming obsessed and being taken over by them. On the third floor are your attitudes and ways of thinking about relationships—what you can expect from others and what you are entitled to. At the tip-top is the piece your partner and other loved ones see—your actual behavior.

This is fine as a metaphor, as far as it goes. But a relationship is, we've finally recognized, a dynamic interaction. Once another person comes into the picture, I prefer the metaphor of the dance to capture the reality of a love relationship. How your lover sways and bends and responds to your cues affects all the elements that make up your experience of love and loving, just as you influence his or hers.

Some attachment styles aren't very compatible. For example, relationships between two avoidant people don't "take," for obvious reasons; both partners are determined to reject emotional involvement. Two highly anxious people don't pair up very well, either; they are too labile and absorbed by their own worries. A very common pairing has one anxious and one avoidant partner. This combination, though problematic, can work; the avoidant partner will be responsive at times, and this reassures the anxious partner at least for a while. Partnerships in which one person is secure and the other is somewhat anxious or avoidant also can be positive; the secure partner offers soothing reassurance to the anxious and an undemanding attitude to the avoidant. Matches between two secures tend to be the most satisfying and stable, since both partners are able to be emotionally available and responsive.

One of the most fascinating discoveries of the past few years is that while attachment styles tend to be stable, they are not immutable. Your personal style can modify your partner's, and your partner's can modify your own. For example, an anxious woman who pairs with a secure man who is consistently open and responsive can learn new steps in the dance of love. Romantic love can change us. With the right partner, we can become more open and more secure. Falling in love can give us the chance to revise our childhood model.

Marcie grew up with a philandering father. As a result, she has shut out potential lovers because she "knows" that they can't be counted on. She has shut down her own longing for connection. But then she finds herself being courted by Jim, an open, loving man. He slowly shows her how to take the risk of trusting; she slowly drops her avoidant strategy and moves into a more secure attachment style. For Marcie, her love relationship with Jim is not just a source of happiness but also a regenerating force that transforms her world and herself.

LOSING AND REGAINING LOVE

Some of us are lucky enough to have been given by our parents a model of what secure love and loving looks like. It is then easier to reproduce it. But some of us, like Marcie, have to trust our instincts and learn all this from scratch with our adult lovers.

Either way, all relationships fall into conflict or distress at some point, and the bond between partners begins to unravel. Given how little we have understood about love and bonding, it is amazing to me how many of us end up creating positive relationships and just how long and hard we fight to try to repair relationships that are floundering.

Knowing how attachment works means that we are not in foreign territory when we find ourselves estranged from or enraged by the person we were convinced was the One and we now see as a Stranger, even the Enemy. We can understand that what we're dealing with is the panic and pain of separation distress, and that we experience it in the same way children do. Feeling rejected and abandoned, we reach out, pursue, and cling with the same anger and despair. Bowlby reminds us that in love relationships, "presence and absence are relative terms." He points out that a loved one can be physically present but emotionally absent. Both as children and adults, we need a readily accessible and responsive loved one to feel secure in our bond. This point is captured in a common exchange between lovers: "I am here, aren't I? Don't I do things for you?" "Then why do I feel so alone?"

Separation distress usually proceeds through four steps: The first is *anger* and *protest*. Little four-year-old Sarah demands, "Don't go away, Mommy. You come here!" Grown up thirty-two-year-old Sarah attacks her husband, saying, "Do you really have to go see your mother, Peter, just when I am so overwhelmed with the kids? You're always working. You never talk to me. You're just selfish. Sometimes I think that you don't need me at all!" In adult relationships, the overt anger can make it hard for a partner to hear the very real underlying anguish. What the partner hears is the criticism and hostility; the reaction is often to turn away in self-protection.

The next step is *clinging* and *seeking*. Little Sarah might say, "I want you to pick me up. I don't want to play. I want to stay here in your lap." The adult Sarah tells her husband, "I have asked you to come home early a thousand times. But right now, right now, you are not even listening to me. You say you love me, but you never hold me. I want you to hold me." And then she cries. If he responds coldly—"Well, you have a funny way of asking. You are always angry with me. Who can listen to that all the time?"—her misery deepens.

The third step is marked by *depression* and *despair*. Adult Sarah at this point may flip into a rage and threaten to leave her husband in an attempt to get him to respond to her, or she may withdraw into a sense of helplessness, the main symptom of depression. In either case, people at this stage, like Sarah, are beginning to let go of their longing for connection and move into grieving.

The final step is *detachment*. In this stage a person, whether child or adult, accepts that the relationship is not going to fulfill his or her longings, stops investing in it, and decides just to let it die. In thirty years of practice, I have never seen anyone come back from detachment.

We must not underestimate the naked force of separation distress. It is wired into our brains by thousands of years of evolution.

Loss of contact with a protective attachment figure once meant certain death. Neuroscientist Jaak Panksepp of Washington State University has shown that mammals have special pathways in their brains dedicated to registering the "primal panic" that results from the loss, even if only momentary, of an attachment figure. This panic is precipitated by any threat of rejection or abandonment (I'll talk more about this in the next chapter).

In a positive relationship, one in which partners have some level of mutual secure connection, this sequence of events can be halted early on. If Sarah and Peter are just going through a rough patch, the protests of step 1 will work. Sarah might say, "Peter, I know I am being very critical of you these days. I don't mean to be. I know you're under a lot of pressure at work. But I am pretty lonely. And I feel scared about us. Where is our closeness? Don't you miss it? I do. I need you to turn to me, even if it's just for a moment or two each day, so that I know I matter to you. Is that possible, please?" To send this kind of protest message, Sarah has to first tamp down her anger, and then clearly state her fears and needs. If Peter hears her and responds with comfort and support, then they can quickly heal their rift. It's just a nick in the bond between them.

Other relationships, however, need professional help. EFT's program of relationship repair builds on the science of attachment. We help couples grasp the survival significance of a love relationship. We help them see all the moves that are triggering their dance of disconnection. We slow down the steps that are taking them into separation and pushing them into panic, and we help them come together to halt this destructive sequence (you'll read more about this in Chapter 7).

But to repair a bond and shape a safe-haven relationship, we have to do more than simply stop creating distance. We have to do what securely attached dyads do naturally: we have to learn to turn toward each other and reveal our fears and longings. This is, admittedly, hard to do, particularly if we are ashamed or don't have the words to express our needs. Words order emotions and thoughts, make them more tangible and workable. EFT helps couples over these hurdles.

One of the finest moments for me is when partners finally disclose their worries and desires and engage with each other tenderly and compassionately. This Hold Me Tight conversation (discussed in Chapter 8) is a transformative experience, for couples and for me. We've recently completed a study of 32 couples and have found that EFT not only helps couples become more satisfied but also can change the bond between partners, making them more securely attached to each other. This is the first time ever that couple therapy has been proven to have this effect.

In my sessions with couples, I see love coming to life. I see it blossoming! These couples have found their way back to the emotional closeness and responsiveness that are the essence of a secure bond. They've regained or perhaps found a new sense of safety and trust with each other. Together they can go on to master the everyday problems of living and enjoy their future as safe-haven lovers.

Attachment theory and the subsequent twenty years of studies on adult bonding are the foundation of the revolutionary new science of love relationships. But significant insights and contributions have come from many other areas as well, including philosophy, biology, ethology, neuroscience, and social-science disciplines, like clinical psychology (you'll learn about them throughout these pages). They offer different notes and harmonies, but together they are creating a new symphony. In his book *Consilience: The Unity of Knowledge,* biologist Edward O. Wilson notes that when he realized "the world is orderly and can be explained by a small number of natural laws," he experienced "The Enchantment." To finally lay out the laws of love and loving brings an enchantment all of its own. And this new science casts a greater spell than any of our earlier visions of romantic love ever have.

EXPERIMENT 1

Think back to the time when you were growing up. Who was the main person you would go to for comfort? Who offered you a secure base from which you could go out and explore? What was the most important thing this person gave you, taught you? If you did not have this kind of relationship, how did you cope as you grew up? Do you have this kind of relationship now, as an adult?

Adult attachment researchers have identified three basic attachment styles, or habitual strategies. Which statement below best describes you?

1. Secure: I find it relatively easy to get close to others and am comfortable depending on them and having them depend on me. I don't worry about being abandoned or about someone getting too close to me.
2. Anxious: I find that others are reluctant to get as close as I would like. I often worry that my partner doesn't really love me or won't want to stay with me. I want to get very close to my partner, and this sometimes scares people away.
3. Avoidant: I am uncomfortable being close to others; I find it difficult to trust them completely, difficult to allow myself to depend on them. Often others want me to be more intimate, but I am nervous when anyone gets too close.

How do you think your own attachment style—your "how to" protocol for engaging with loved ones—affects your love life?

EXPERIMENT 2

Adult bonds are more reciprocal than parent-child bonds. Think of one thing your partner does that makes you feel precious and loved.

How do you reciprocate—that is, what do you do to make him or her feel the same way?

Do you *know* specifically what makes your partner feel precious and loved?

If not, can you ask?

EXPERIMENT 3

When you face a recurring event that makes you anxious, such as getting on a plane, giving a speech, or being evaluated by your boss, which loved one from your past pops up in your mind?

Do you see this person's image, hear his or her voice, remember some soothing words? Can you use this memory to calm yourself and regain your emotional equilibrium?

See if you can write down the message this person conveys and how it helps you change the way you view the situation

For example, Amelia gets nervous when she goes to the dentist. As a child, she once passed out in the dentist's chair, and since then she has dreaded her teeth-cleaning appointments. But, she says, "I remember my dad always telling me how strong I am and how, even when bad things happen, I will come through, I can cope. I see his face, and his smile tells me how much he believes in me. Then it's okay; I can tolerate the visit."

PART TWO

The New Science of Love

CHAPTER 3

The Emotions

The emotions do not deserve being put into opposi-
tion with "intelligence." The emotions are themselves
a higher order of intelligence.

—O. Hobart Mowrer

Strong emotion is the essence of love—and strong emotion
is what has given love a bad rap. We don't understand intense emotion, and we don't trust it. We want the joy and
elation love brings. They lift us up out of our dull, mundane routines and make us feel alive and significant. But we abhor the fear
and anger and sadness that also attend love. They drop us into deep
pits of desolation and despair and make us feel helpless and out of
control.

Colin tells me, "In the beginning, in the first infatuation, feeling
like I was being swept away was intoxicating, thrilling, even. The
excitement was so high. I felt so alive. This is what I had longed
for all my life. All the stupid sentimental songs suddenly seemed

so true. I took emotional risks without even thinking about it. But then Donna's old boyfriend came back into town, and she met him for coffee. Even though I agreed to their meeting, suddenly everything seemed different. Waiting for her outside a restaurant, standing in the rain, I realized that I was out on some kind of limb here. I wasn't in charge of what was happening at all. Suddenly I felt so vulnerable. I didn't know whether to run or rage. When Donna arrived, I stayed cool. I told her that I was going to be busy for the next few weeks and I cut our date short."

As Freud remarked many years ago, "We are never so vulnerable as when we love." If we don't understand the intense emotion that love engenders, then love will always be a scary proposition. Thankfully, a radical new view of emotion and its role in love relationships has been emerging. In the past two decades, nearly every "fact" about emotion that was drummed into my head in grad school has been repudiated. We owe this largely to advances in technology. We no longer have to rely solely on patients with brain anomalies or terrible head injuries—like Phineas Gage, the 19th-century railroad foreman whose personality changed after an explosion rammed an iron bar through his cheek into the emotional center of his brain—as subjects for study. With the fMRI scanner, we can look inside normal brains and actually see in real time where emotion arises and how it operates. And what we've learned is astonishing.

Technology has given the lie to long-held assumptions about emotion as a random, irrational impulse. Emotion, we've discovered, is a sharp, smart force that organizes and elevates our lives. It is what transforms existence into experience. "I do not literally paint that table but the emotion it produces upon me," observed Matisse. Emotion is what turns an object into a memento, an event into a happening, and a person into the love of your life. Nor is emotion a selfish, corrupting drive leading inevitably to destructive excess and devilish sins, as my first teachers, Catholic nuns, warned me. We've now learned that emotion is, in fact, the foundation of key elements in civilized society, including moral judgment and empathy. To feel for someone is the root of caring action.

Equally amazing is what the new research reveals about the impact of emotion in our closest relationships. The message touted by popular media and therapists has been that we're supposed to be in total control of our emotions before we turn to others. Love yourself first, and then another will love you. Our new knowledge stands that message on its head. "For humans," says psychologist Ed Tronick of the University of Massachusetts, "the maintenance of [emotional balance] is a dyadic collaborative process." In other words, we are designed to deal with emotion in concert with another person—not by ourselves.

Love relationships aren't meant only to be joyrides; they're also restorative and balancing meeting places where negative emotions are calmed and regulated. It's a little like the old adage "Two hearts are better than one"; indeed they are. When we find what Harry Harlow called "contact comfort" by moving close to another person, the impact of every risk or threat is reduced. In horror movies, the hero or heroine is always alone when the ghoul or monster first appears but finally triumphs over fear and fiend with the help of a buddy.

In fact, the reason that distress in a relationship so often plunges us into inner chaos is because our hearts and brains are set up to use our partners to help us find our balance in the midst of distress and fear. If they instead become a source of distress, then we are doubly bereft and vulnerable. As Terry tells his wife, "That you would do this to me, you of all people. The one I count on. I am so confused. If I can't trust you, who can I trust? I thought you had my back; you were my safe place; but now it seems like you are the enemy, and there is no safety anywhere." The other side of the coin is that loving connection is the natural antidote to fear and pain.

Jim Coan at the University of Virginia, one of the most creative scholars in the new field of social neuroscience, put women, all happily married, in an fMRI machine and took pictures of their brains as they saw small circles and x's flash in front of their eyes. They were told that when they saw the x's, there was a 20 percent chance that they would receive an electric shock on their ankles. After each shock, they rated, on a simple scale, how much it hurt. The twist in the experiment: sometimes the women faced the shock threat alone; other times they were with a stranger who came into the room and held their hand; and still other times they faced the shock threat with their husband clasping their hand.

The results were fascinating. When the x's popped up in front of their eyes and they were all alone, their brains lit up with activity like a Christmas tree. Alarm was everywhere. And they rated the shock, when it came, as very painful. When the stranger grasped their hand, their brain reacted with less alarm to the x's, and they found the shock less painful. Isolation is traumatizing and exacerbates our perception of threat. But what was really interesting was that when their husband gripped their hand, their brain barely responded to the x's (just as if someone had told them it was raining outside), and they said that the shock was simply uncomfortable. This is love in action, offering us safe harbor, a place to calm our terrors and find equilibrium.

Recently Jim and I did a variation of this study with women in distressed marriages who were in therapy. They and their partners reported on questionnaires that they were insecurely attached and were undergoing EFT to improve their relationships. Before therapy, the women reacted just as the women in the original study did: their brain lit up in alarm and pain, and the shock really hurt. A stranger holding their hand eased their fear a little. But clasping their husband's hand had little or no effect. The spouse was not a safety cue in these insecure and troubled marriages.

After twenty sessions of EFT, however, the women were happier and more secure in their relationships, and when they saw the x's and had their husband's hand to hold, their alarm response was virtually eradicated and their pain was judged "uncomfortable." What was especially striking was that their prefrontal cortex, where emotions are regulated and controlled, did not even blip. With the presence of a mate with whom they felt more securely bonded, the women were not just able to cope differently with the pain, they registered even the threat of being shocked differently. This marks the first time that a systematic intervention aimed at changing interaction with a loved one has been shown to have an impact on the brain. This means that, with the right kind of therapy, we can begin to create a safe-haven relationship.

Learning to love and be loved is, in effect, about learning to tune in to our emotions so that we know what we need from a partner and expressing those desires openly, in a way that evokes sympathy and support from him or her. When this support helps us balance our emotions—staying in touch with but not being flooded by them—we can then tune in to and sensitively respond to our partner in return. We can see this in movies of moms and secure kids in the Strange Situation experiment, and we see it in our research tapes of adults in therapy who succeed in mending their relationships. In these moments, we are what John Bowlby called "effectively dependent"; we can call to others and respond to their call in a way that makes us and our connection with them stronger. Once we are balanced, we can turn to the world and move in it with flexibility, open to learning and able to look at the choices available to us in any situation. Nothing makes us stronger and happier than loving, stable long-term bonds with others.

WHAT IS EMOTION?

Distrust of emotion has long been a hallmark of Western civilization. It dates back at least to the days of ancient Greece, when Stoic philosophers argued that the passions, love included, were destructive and had to be checked by intellect and morals. Down through the years, emotion has been viewed primarily as an attribute of our base animal nature, crude and sensate. After all, we "feel" emotion; it is a visceral force. Reason, in contrast, removed from the body and residing "in the head," has been viewed as evolutionarily superior, a reflection of our higher spiritual self. We must rise above emotion if we are to be a truly civilized society. Social critic Marya Mannes said it succinctly: "The sign of an intelligent people is their ability to control emotions by the application of reason."

The case against emotion seems to stem from two factors: its unstoppable power—indeed, it can overtake us in less than a second—and its apparent randomness and lack of logic. Research now paints a much different view. Emotion is actually nature's exquisitely efficient information-processing and signaling system, designed to rapidly reorganize behavior in the interests of survival.

Emotion apprises us that something vital to our welfare is occurring. We are bombarded by hundreds of thousands of stimuli every second of every day. Emotion automatically and reflexively sorts through the barrage, picking out what matters and steering us to the appropriate action. Our feelings guide us in issues large and small; they tell us what we want, what our preferences are, and what we need. We choose pistachio ice cream rather than vanilla because we have a better feeling about it. Research with brain-damaged people shows that without emotion to guide us, we have no compass. We are bereft of direction and have nothing to move us toward one option rather than another. We are stuck pondering all the possibilities.

Emotion is the great motivator. It comes whether we will it to or not, and it stirs, even compels us to act. The word *emotion* derives from the Latin *movere,* meaning "to move out." We see its power most clearly when we sense we are in immediate physical danger. If we're charged by a rabid dog or a rampaging rhino, we feel fear and make tracks in the opposite direction. Charles Darwin, the first scientist to point out emotion's survival value, would regularly visit the London zoo to stand in front of the puff adder cage. He knew that staring at the snake eye to eye would make it strike out. He also knew, as a reasoning being, that he was perfectly safe, since the adder was behind glass. Darwin stared, determined not to flinch, but no matter how many times he tested himself, when the reptile attacked he jumped back.

Emotion can spur us to act even when survival does not appear to be an immediate issue. During 9/11, a woman named Julie was at work in the South Tower of the World Trade Center when the first plane hit the North Tower. Instructions over loudspeakers told her and her colleagues to stay put in their eightieth-floor offices. But overwhelmed by fear, she headed down the stairs. She had reached the sixty-first floor when the second plane hit her building. She made it home. Of course, emotion is not an infallible alarm system, as Darwin's experience demonstrates. Julie could have made the hot and anxious trek down eighty flights for nothing. But in survival, false positives are always more valuable than false negatives. You're better off heeding a warning emotion than ignoring it. As George Santayana pointed out, it is often "wisdom to believe the heart."

Emotion is also the great communicator. It swirls within our bodies and flows out, whether we want it to or not, as signals to others. It spurs our own behavior and conveys our deepest needs to others as well as theirs to us. As such, it is vital to our love relationships. Our partners are central to our sense of safety. How can they shelter us, be our safe harbor, if they don't know what we are afraid of and what we yearn and hunger for? Emotion is the music of the dance between lovers; it tells us where to put our feet, and tells our partners where we need them to put theirs.

We broadcast emotion mainly through our facial expressions and tone of voice, and we apprehend and comprehend these signals instantaneously. It takes just one hundred milliseconds for our brain to register the smallest alteration in another person's face and just three hundred milliseconds more to feel in our own body what we see in that face—to mirror the change we see (I will talk about just how this mirroring process occurs in the next chapter). Emotion is contagious; we literally "catch" each other's sentiments and feel what the other person is feeling, and this is the basis of empathy.

NAME THAT EMOTION

Dorothy Parker once famously panned Katharine Hepburn's performance in a play as "running the gamut of emotions from A to B." We know that there are more emotions than that—many more, most people would say. But how many? Some twenty years ago, I read a book for therapists claiming that there were thirty-nine basic feelings. However, most social scientists today agree that there are only *six* innate and universal emotions: fear, anger, happiness or joy, sadness, surprise, and shame (some theorists divide shame into disgust and guilt). Each one leads naturally to an action. In anger, we approach a challenge or a frustration; in surprise, we pay attention and explore; in fear, we freeze or flee. The fact that negative emotion predominates in this list speaks to the existential significance of emotion. In survival terms (as reporters know), bad news and negatives are more important. You've probably noticed that love isn't on the basics list, and I'll go into that a little later.

The list derives from American psychologist Paul Ekman's pioneering research into facial expressions. Darwin believed that emotional display is biologically determined and universal across societies. But in the 1950s, another view held sway: that emotion's manifestation is culturally dictated and learned. Which was right? Ekman's evidence favors Darwin: he has found that people in literate cultures of the West and East agree on the nature of the emotion they see in pictures of faces. People from all over the planet can, in fact, read certain key emotional expressions and assign the same meaning to them.

Ekman first traveled in the 1960s to the highlands of New Guinea to meet with a tribe, the Fore, that has no written language. An isolated people, they had never seen movies or television. Ekman showed them pictures of Westerners' faces displaying different emotions and, through an interpreter, asked them, "What is happening to this person, and what is going to happen next?" He found that they could pinpoint what the Western person was feeling and predict his intentions. He visited the tribe a second time and told them stories, then asked them to match the story with a picture of a facial expression. They easily did so. Ekman also took photos of tribesmen with varied facial displays, and when back in the United States he asked college students to interpret the emotional content. The students identified the tribesmen's emotions and how this linked to their intentions.

Ekman's findings have been replicated by colleagues studying other isolated groups. The conclusion they have all reached is that the display of the six basic emotions and the ability to assign the same specific meaning to each cuts across cultures. In other words, there is a universal language of emotion. In anger, the eyes widen and stare, the brows contract, and the lips compress. These expressions do not need to be learned; the congenitally blind also show them. Culture appears to be significant in one respect, however: it influences which facial features we focus on. For example, if we're from the West, we pay more attention to the mouth and brows; if we're from the East, we concentrate on the eyes.

Just knowing that there are basic emotions we all feel and recognize can make a huge difference in our everyday lives. Psychologist Matthew Lieberman of the University of California, Los Angeles, has demonstrated that the simple act of naming an emotion calms the emotional center of the brain. In an fMRI study, Lieberman showed people images of faces with negative expressions—for example, grimacing in anger. When subjects were asked to identify the sex of the person in the image, the emotional area remained highly activated. But when they were asked to label the feeling they saw by choosing between two words written under the image, their brains calmed down. Naming an emotion begins the process of regulating it and reflecting on it.

I see this happening in therapy. As Bernice tells me, "Well, I guess if I pay attention instead of going off in my head, I can see that my husband is sad right now. Usually I just freak out—get all confused and dithery. It's silly, but it feels better to recognize that he is sad. It's kind of like pinning everything down; the cues he is sending and my inner responses make sense then. Everything seems clearer, more manageable. Last time, I didn't just clam up. I was able to tell him, 'When you get so sad, I don't know what to do, and that scares me. I think you are getting depressed again.'" What we name we can tame; when we give meaning to something, we can tolerate it and even change its impact.

So what about love—why isn't it on the basics list? A few of my colleagues say it should be, but I don't. Love doesn't have a distinct facial expression. It's not a single emotion, a lone note. It's a mix of feelings, a medley. In point of fact, it's a state of being that encompasses all the basic emotions. When we love, we can be joyful, sad, angry, afraid, surprised, or ashamed—often at the same time. Writer Jeffrey Eugenides puts it beautifully: "Emotions, in my experience, aren't covered by single words. I don't believe in 'sadness,' 'joy,' or 'regret'...It oversimplifies feeling. I'd like to have at my disposal complicated hybrid emotions, Germanic train-car constructions like, say, 'the happiness that attends disaster.' Or: 'the disappointment of sleeping with one's fantasy.'"

GENERATING EMOTION

Today, we have not only identified the main emotions, we also know how they are generated and processed. First, there is a trigger or cue—say, a beautiful sunset or a frown on your lover's face. This cue is picked up by the thalamus, a structure deep in the brain, and given a fast read to identify which emotion is called for and ready the body to react. The information is then relayed onward. If the initial rough assessment is that immediate action is required—as when, for example, you're being attacked by an intruder with a knife and your life is at stake, the message goes straight to the amygdala, a small, almond-shaped organ between the temporal lobes. If there is no such urgency, information travels on a more circuitous route from the thalamus to the frontal cortex before heading to the amygdala. The cortex is the thinking part of the brain; it assesses the exact meaning and significance of the stimulus, but this assessment is too slow to be useful in critical situations. Finally, a compelling action emerges, and the body responds. In anger, blood is directed to the hands to prepare us to fight; in fear, it is directed to the feet, to prepare us to flee. The entire sequence occurs without our being aware of it; it is swift and supremely logical.

The idea that emotion also involves reason will surprise most people. In the past, emotions were believed to originate strictly in the right brain (the "feeling" side), and thoughts were believed to originate in the left brain (the "rational" side). As one scientist wrote, intense emotion involves "a complete loss of cerebral control" that contains no "trace of consciousness." Today, we have a much more nuanced, integrated picture. There is evidence that the right brain is more active when we feel highly arousing emotions, like anger. But we now know that such emotions, whether positive or negative, generally activate both sides of the brain and that the frontal cortex, once considered the exclusive province of reason, processes emotional cues.

Dividing the brain into parts, and separating emotion from reason, is illusory. A functioning brain is an integrated brain. All parts work together to create our experience at any moment. Interconnection and interdependence are the name of the game—in the brain and in relationships.

One of the insidious effects of the cult of independence in the Western world is that we've been taught that suppressing our negative emotions is an effective—indeed, the optimal—strategy for getting along in life. I remember learning early on the adage "If you can't say anything nice, don't say anything at all." My clients tell me that they put huge effort into holding back when their partner upsets them. But that, we've now discovered, often simply exacerbates relationship difficulties. Moreover, repression takes a huge physiological toll.

Psychologist James Gross of Stanford University has directed a series of fMRI experiments assessing the effect of suppressing emotions compared with another strategy, reappraisal—that is, changing the way we evaluate an emotional situation. One of the most interesting studies asked 17 women (women are considered more emotionally expressive than men) to watch fifteen-second film clips of either emotionally neutral nature scenes or "disgusting" events, including vomiting, surgical procedures, and animal slaughter. The women were instructed either to try to hold back their reactions to the repulsive clips ("keep face still") or to try to reappraise the events by adopting a more general perspective, like that of a medical professional watching the film.

Scans showed that suppression actually heightened activity in the amygdala, "fear central" in the brain. Stifling emotional reaction had a rebound effect: the women became so stressed and tense from holding back that the negative emotional effect of the disgusting clips was exacerbated. By contrast, the more distancing reappraisal strategy reduced the women's negative emotional experience of the disgusting films. Scans showed activation of the prefrontal cortex, the region of the brain that regulates emotion and turns down activity in the amygdala.

Why is reappraisal a more effective strategy? Reactive emotion flashes up very fast. Reappraisal intervenes as emotion is being generated and thus is able to modify and shape it. Suppression, on the other hand, occurs after the emotion registers in the brain. We have to work very hard to push down intense emotion; our heart rate speeds up, and stress chemicals pour out. Think of capping a ready-to-erupt volcano: the bottled-up force makes the eventual explosion stronger. That's why we see people suppress, suppress, suppress, then blow!

Smothering emotion is bad not just for us but also for our love relationships. The effort is exhausting and distracts us from attending to emotional cues coming from our partner, curtailing our ability to respond. James Gross has shown, too, that the tension created by suppression is contagious: our partner picks up the strain, and becomes stressed as well.

The most functional way to regulate difficult emotions in love relationships is to share them. We know that confiding helps us reorganize our thoughts and responses, get clear about our priorities, receive new information and feedback, and feel comforted and calmed. The complicating issue is that the partner we share with is also often the trigger for our bad feeling.

FEAR AND LOVE

Most folks tend to associate love with the emotion of happiness. When people in studies are given lists of words and told to group them into categories, they generally place love under "joy." But to scientists who study love relationships, the most fascinating emotion is fear. Fear is the most powerful of all the emotions. Not surprising, since it is our basic survival mechanism, an alarm that blares when threat looms and that prompts us to escape.

Psychologist Mario Mikulincer, perhaps the most prolific researcher on attachment in the world, speaks to me in a soft, musical voice from his lab in Herzliya: "I am a Jew, and with the history of my people, I became fascinated with how people deal with fear, helplessness, and lack of control," he says. "Since we knew from research into children and their caregivers that feeling securely attached increases one's sense of mastery and helps modulate negative emotions, we decided to examine how our sense of attachment impacts our fear of death. And we found that securely attached people seem to be less afraid. Anxiously attached people's fears center around not mattering to anyone anymore and leaving others. Avoidant people's fears focus on the unknown nature of death. I was captivated. I realized that our bonds with others are not only our most crucial source of vitality but our strongest defense. Suddenly I realized that I was studying the power of love!"

Fear clangs noisily, too, when our love relationships, the main source of our emotional support and comfort, seem to be in jeopardy. More and more evidence is emerging that the nexus for the social brain is the amygdala, the main processing site for fear. Jaak Panksepp of Washington State University has been studying the brains of rats for thirty years. In structure, rodent brains are surprisingly similar to ours. Panksepp's work reveals that rats who bond with their mates and rear their young with care have a specific neural pathway in the amygdala that switches on automatically when a loved one is suddenly perceived as unavailable, such as when their mate is temporarily removed from their side. Panksepp has shown that this separation plunges them into what he calls a "primal panic."

Panksepp is convinced that a comparable pathway exists in the brain of all mammals who form close ties with others, including humans. I am convinced of this, too. "When Michele just turns away and shuts me out like I don't matter at all to her, I go into some kind of meltdown," says Darren in my office. "Does this mean I am crazy?"

"No," I assure him. "It means you are a mammal in love who suddenly senses a lack of connection. Your brain takes this as a danger cue. It codes it as a threat to your safety and well-being."

In fear, muscles tense, stress hormones release, blood rushes, thoughts of pain and other harm arise, and the impulse to freeze or flee forms. The elements of this experience are inescapable and unfurl predictably. Each element inexorably evokes the next, and the more times it is laid down in the neural circuitry, like a track repeatedly run over in snow, the more automatic the entire sequence becomes.

Andrew grew up with volatile, abusive parents and is sensitized to loud voices. So when his wife, Amy, raises her voice, he moves into fear faster and more intensely than someone who was reared in a tranquil, supportive home. "Deep down, I am so wary and so vulnerable," Andrew confesses to Amy during couple therapy. "I am always ready to run away. It is hard for me to let you in. I always assume the worst is happening. I guess I need lots of reassurance that you do want to be with me, and I need for you to be patient as I learn to trust."

Unhappy partners often are visibly angry, but the anger is usually secondary to a deeper sense of fear. Emma reminds Tim that

they have a special date on the weekend to celebrate their ninth anniversary. Tim shrugs and comments that they will have to change
it; he promised to attend a party with his boss. Emma explodes in
anger. But if we were to freeze-frame this encounter, we'd see that
Emma's first emotion when Tim announced the cancellation was
fear. If she were able to slow down and pay attention to her fear that
she is becoming less important to her husband, her action might be
very different. Instead of erupting angrily, she might ask for reassurance. But Emma does not register this anxiety. When she talks
about this row in my office, she looks angry, and she accuses Tim
of selfishness. Emma's outburst in turn triggers her husband's fear
of failure and rejection. He becomes still and silent. This response,
unfortunately, reinforces Emma's fear. Their different ways of dealing with their emotions become part of a script, a pattern in their
marriage. If the script this couple is following becomes fixed, the
relationship is in trouble.

To reiterate this in a more general way: the way we regulate
and process our emotions becomes our habitual way of signaling
and engaging with others. It becomes our social script. The more
narrow the focus of the script, the more limited are our ways of
dancing with others.

PAIN IN LOVE

Later, when Emma feels safer with Tim, they discuss the above
incident, and she is able to broaden her focus and explore her experience, acknowledging that the "hard" emotion she showed him
was not the whole picture. It was the "soft" emotion of hurt that
was the main music playing for her in their conflict. Some have
suggested that hurt should be included in the list of basic emotions. But we now know from various studies that this kind of hurt
is a composite emotion, made up of, on the surface, anger; on a

deeper level, sadness at a sensed loss of feeling valued by another; and, on the deepest level, fear of rejection and abandonment. As Emma shows her hurt, it changes the script with her husband. It prompts tenderness in him and reassures him that he is valued by his wife.

Just as we have not understood the role specific fears play in love, we are only now understanding the tangible physical nature of social or relational pain. Until recently any parallel between emotional pain, such as rejection, and physical pain, such as burning your arm, was thought to be caused purely by overlapping psychological distress rather than by any shared sensory processing. In fact we often downplay others' hurt by comparing it to the "real" hurt of physical injury. Amanda says to Roy, "You act like I stabbed you just 'cause I got a little critical. Don't you think you are being a little melodramatic here?"

It is now clear that there is a literal neural overlap in the way we process and experience relational and physical pain. Both pains, as experiments by psychologist Naomi Eisenberger of UCLA attest, are alarm systems, designed to grab our attention and focus our resources on minimizing threat. The threat in hurt feelings, arising from triggers such as rejection by a loved one, is emotional loss and separation. In mammals, perhaps because of their need for extended maternal care, isolation is a clear danger cue: it registers as a physical threat to survival.

Eisenberger and her colleagues arranged for subjects, while lying in a brain scanner, to play Cyberball (a virtual ball-toss game) with, so they believed, two other players. In fact, they were playing with a computer programmed to act as if the other players were deliberately refusing them the ball. The subjects reported feeling excluded and ignored, and their brain scans revealed significant activity in the anterior cingulate cortex (ACC), the same region that registers physical pain.

This neural overlap explains why, as researchers have found, Tylenol can reduce hurt feelings and emotional support can lessen physical pain (including that of childbirth, cancer treatment, and heart surgery). Our need for connection with others has shaped our neural makeup and the structure of our emotional life.

SHAPING RELATIONSHIPS

We learn about the nature of emotion and what to do with it in our first attachments. If we are lucky, over the course of thousands of interactions with loved ones who are exquisitely responsive to us, we learn to tune in to, order, and trust our emotions and those of others. We can also use the supportive responses of those closest to us to shape and modify our own emotional reactions. Good relationships in childhood do not mean that our emotional life becomes consistently even and positive, but it does mean that we are more likely to discover that our negative emotions are workable and useful and that our positive emotions can be trusted and rejoiced in.

The good news is that even if we were emotionally starved in our childhood relationships, our adult lovers offer us a second chance to learn new and more effective ways to deal with our emotions and signal our longings to others. At the end of the process of EFT, Marion, who was physically and sexually abused by the people she depended on as a child, tells me, "It's a strange thing: I have had these inner demons, these terrible fears about myself all these years. I could never risk letting anyone see me. It's like I would get hijacked by terror if anyone got really close. If I trusted them and they hurt me again...it felt too risky. But now, with Terry, I can touch my shame and my fear, and ask for his help with these feelings. And when he gives it, I calm down and I feel reassured and somehow more whole. It's like a loop—more safe connection leads to more feelings of safety inside and vice versa." More secure bonding teaches us how to tolerate, work with, and use our emotions, and being able to manage our emotions in turn helps us adapt to and connect with others.

A secure relationship is one in which we learn to become emotionally intelligent. Loving partners help us when we are confused and unsure about our feelings, as when we feel too little or too much. When we feel too little, we say things like, "I don't know how I feel. Maybe I feel sad, but I don't know why." We cannot order our experience into a coherent whole; we cannot find the direction in the emotion. Sometimes we feel "flat" or cut off from our emotions altogether. The inability to touch or name emotions leaves us aimless, without an internal compass to steer us toward what we need. The inability to show emotion also completely leaves our lovers hanging in space. No signal, no music, no dance, no relationship.

On the other hand, plugging in to too much emotion can be overwhelming and chaotic. I can remember being shocked by the way my grief at my mother's sudden death took over my body and my world. As a client remarked to me, "Grief is like drowning in a bottomless sea." At such moments we are all too aware of how fragile we are. People use images to capture experiences of overpowering emotion; for example, my clients use phrases such as: "To face the fear of reaching for someone is like walking through fire for me"; "His anger hits me like a Mack truck. I am knocked down, flattened"; "The shame hits me like a wind—so, so cold. Suddenly I am helpless. All I can do is hunker down and disappear." We seem to be able to capture emotions best in images. They bring together perfectly the elements in the experience of emotion: the triggers, sensations, meanings, and the urge to act.

If we find ourselves caught in the too-much-or-too-little mode across lots of situations and relationships, chances are that we are having a problem with emotional balance, with regulating our emotions. The ability to find this balance is the most basic lesson we learn (or not!) from our early attachment figures. Those of us who have had even just one such positive relationship with a parental figure gain an advantage: we acquire a procedural map of how to hold on to our emotional equilibrium and connect with others. Being in balance allows us to move in many directions easily and thus have more ways of responding to and dancing with others.

When we are emotionally poised—either because that is our personal style or because we are tightly connected to another—we are less triggered. We do not hold on to and expand on any fear of rejection or betrayal kicked up by small slights and injuries. If we do feel hurt, we have more faith that we can share those feelings and get our lover to respond in a way that heals us. We are not flooded with alarm messages from our body or swamped by catastrophic thoughts; we can listen to our longings and risk asking for help in retrieving our balance. All this adds up to the fact that the more secure we are, the more able we are to turn emotion up or down with relative ease. A secure base creates safety that continues to foster personal growth, emotional balance, and loving connection. Being able to securely attach is the gift that keeps on giving!

HAPPINESS

We tend to focus on negative emotion because cues relevant to survival are given priority by our mind and body. But positive emotion is a powerful force as well. Life is, after all, a constant search for just this! Studies now show that happiness is not only a sign of flourishing but also the impulse that creates well-being. Just as sunlight makes gardens grow, joy makes us more alive and adventurous. It moves us forward and outward, pushing us to explore novel objects and places and engage with loved ones and strangers. In psychological terms, it sends us into "approach" behavior—but in a softer, more inquisitive way than does anger, which has a harder, more assertive quality. Negative emotions, such as anger and fear, narrow our focus, while positive emotion expands the range of our thoughts and creates the urge to play and experiment.

When we watch children having fun together at the park, we can see this easily. When I watch partners who have repaired their relationship and are now preparing to leave therapy, I see a new web of smiles, touches, and laughter connecting them. They are ready and eager to become more open to each other. Annie beams at Josh and tells him, "You are so funny. I never realized that before. It must be loving me—it's growing your brain." Josh, a strait-laced, introverted man, taps his thumb on his nose, crosses his eyes, and giggles. "Press the button," he says. "More neurons coming up, 'cause I sure do love you." Her eyes fill with happy tears.

So never mind the obvious advantages of joyfulness: if we stick with being stodgily scientific, what exactly does joy do for us, besides make us feel so good? Psychologist Barbara Frederickson of the University of Michigan asked people to view three types of film clips: those depicting situations filled with joy, those filled with fear and anger, or those with a neutral emotional tone. Then she told them to imagine themselves in the scenes. After the screening, viewers were asked: "What would you like to do right now?" They came up with many more responses after viewing the joyful clips—that is, they had a broader "thought-action" repertoire. Even the modest version of joy—contentment—generated more answers to the question. Positive emotions turn on our curiosity and desire to engage and explore. They set us up for openness and learning. Joy, for example, invigorates us.

But this is not all positive emotions do. They also undo negative emotions. We all know that making our partner laugh after we've made a careless, hurtful remark soothes upset feelings and eases the way back into harmony. Great literature is full of this undoing. The war-torn hero, aching with grief, stumbles into a church, is uplifted by the music of the choir, and turns toward life again. Positive emotions remind us at such times that suffering and uncertainty are not the whole story in any human life. Positive emotions and beliefs fuel resilience and help us bounce back from adversity. They generate even more positive emotions in an upward spiral.

This is surely part of the power of love. Love, at its best, brings a cornucopia of good things: joy and contentment, safety and trust, intense interest and involvement, curiosity and openness.

If science has taught us anything about emotion, it is that we should never underestimate its power and value. It has shown us how emotion figures into our most intimate relationships and shapes them. And it has taught us that we can use those relationships to temper our negative feelings, dampen their toxicity, and be inspired by positive emotions to reach out to others and to the world. In his book *The Wise Heart*, Buddhist teacher Jack Kornfield offers a beautiful image for our new understanding: "We can let ourselves be carried by the river of feeling—because we know how to swim."

EXPERIMENT

The better you are at listening to and distilling your emotions and sending clear emotional signals, the better your relationships will be. Science is disciplined observation—forming a hypothesis and testing it. You do it every day.

Sit quietly for a moment with a pen and paper in front of you. Then think about this question: Can you pinpoint a time, either in your current love relationship or in a past relationship, when you felt hurt or scared by the dance you were caught in with your loved one?

See if you can focus on the moment when these feelings crystallized. What was the trigger? Was it a look on his or her face? Was it a word used or a conclusion you drew from the way the dance was moving? Write this down.

See if you can find the trigger—the body sensation, the catastrophic thought about you or the relationship—and the action impulse that appeared with it. Did you want to run, to turn and fight, to crawl under the rug? Write down any of these that you can name.

What did you do? This question is hard. Try to focus just on the action, use a verb, and ignore the desire to defend yourself or prove your partner was wrong.

Can you find a new or a "perfect" word that distills your emotional experience? (A recent fMRI study found that just being able to put feelings into words seems to calm our painful and difficult emotions.)

What do you think your partner saw? Did he or she see what you were actually feeling on a deep level, or just annoyance or blankness? Did you signal your real emotion, or did you throw up a mask to protect yourself?

What do you think will happen if you tell your partner about your deep feelings now? What does this tell you about the state of your relationship?

Your answers will probably depend on how alarmed you were. If you were very apprehensive, the emotional cue traveled the fast road to your amygdala, the processing center for fear. That may make it hard for you to think this through, but you probably will be able to pull up your instinctive reaction. If there was less alarm and urgency, the message went the longer route, through your cortex, where it was thoughtfully assessed, and then on to your amygdala. This path makes it easier to pinpoint your reaction.

Paying attention to the way your emotions unfold in interactions with your partner can reveal important patterns. Once you recognize a sequence, you can exert more control over how you react and offer your partner guidance as to the response you need and want from him or her.

For example, Sally tells John, "When you act tired and don't want to make love, it's okay, I can handle it. Unless you turn away from me in bed and instantly fall asleep. Then I automatically go into this funk of spiraling thoughts: 'I don't exist. He can just turn away. He will leave me, like all the others have. It's just a matter of time. You fool, don't trust him.' Once this happens, I am stuck in anger all the next day. I don't want to go into this panic." John responds by offering to hold her when he is tired, so they can fall asleep together. Sally also agrees to tell him next time she leaps to thoughts of him leaving her.

CHAPTER 4

The Brain

> My own brain is to me the most unaccountable of
> machinery—always buzzing, humming, soaring,
> roaring, diving, and then buried in mud. And why?
> What's this passion for?
>
> *—Virginia Woolf*

You walk into a room—and there he is. He turns and, spotting you, grins, and you light up. Your heart flutters, your fingers tingle, you grin back. You feel no threat; indeed, you feel oddly safe. His face reminds you of your beloved father's. He has the same smile, and, like your father, he seems kind and funny. He also looks a bit like that movie star you lust after, the one with the blue eyes, broad shoulders, and sculpted abs. Hmm; very sexy. You move forward, and so does he. You shake hands, then stand together, chatting. After a while, you begin to mirror the way he stands and moves his hands. When he shifts weight to his left foot, you shift yours to your right. When he crooks his arm and sets it on his hip, your arm soon finds its way to your own hip. He mentions a hassle at work, and you know just what he's feeling. Suddenly, you feel close, connected. You are falling in love.

We feel love in our skin and, we say, in our heart. But as new science is making clear, the true locus of love is the brain. That would have shocked the ancients, who almost uniformly held the brain in low esteem. Egyptians mummifying the dead scrupulously preserved the heart and other organs for use in the afterlife but were so unimpressed with the brain that they routinely scooped it out and threw it away. The Greeks, too, were generally dismissive. Aristotle ruled the brain "an organ of minor importance" whose duty it was to cool the blood. Hundreds of years later, Descartes concluded the brain was a kind of antenna by which the spirit communed with the body.

Today, thanks to new research techniques, we've gained more knowledge about the brain in the past twenty years than we had in all the centuries before. We also know that the three pounds of furrowed, jellylike matter that rests inside our skull is integral to the process of dancing lovingly with another. Indeed, the brain is a profoundly social organ, oriented toward making and managing connection with others. From our earliest days, our brain grows and develops in response to our love relationships, and as we mature, our brain actively works to fasten us to our loved ones. Indeed, says psychologist Dan Stern of the University of Geneva, the brain is so relational that our nervous system is actually "constructed to be captured by the nervous systems of others, so that we can experience others as if from within their skin, as well as from within our own."

LOVE SHAPES THE BRAIN

Our brain thrives on social connection from the day we are born. Our early relationships build the brain, literally. In the first four years of our lives, our brain grows at a very fast pace as emotional interactions with a loving parent or caregiver kindle a host of biochemical processes that boost nerve growth and connectivity. That gelatinous tissue residing in our cranium is actually a collection of one hundred billion neurons, or nerve cells, each of which puts out little tendrils, called dendrites, toward its nearby fellows. These neurons talk to each other by firing electrical and chemical impulses across the gaps, or synapses, between them. Think of neighbors chatting across a backyard fence and you will have the general idea.

Unlike neighborly chats, however, neural signaling happens almost instantaneously and without our ever being aware of it. Moreover, the chatter between neurons is nonstop, critically so. Leave a neuron alone and it dies; give it only an occasional call and it shrinks. This constant dialoging structures our brain. And the more often neurons talk to each other, the easier and stronger the connection becomes. Activation leads to architecture. "Fire together, wire together," as the saying goes.

Emotional interaction advances brain development, and lack of it does the reverse — dendrites don't branch out; the tendrils that relay signals are fewer and stunted, and messenger chemicals are in shorter supply. Infant monkeys who are isolated from their mothers or mother substitutes show gross deficits in multiple areas of the brain, including those involved in the processing of emotion, such as the hippocampus. They display stereotypical behavior, such as repetitive rocking and head banging, and they contract more frequent illnesses of almost every kind. Isolated human babies, such as those reared in institutions, show similar effects. Many sicken and die at an early age. Survivors often mature with attention problems and cognitive and language deficits.

Needless to say, all this affects the ability to form and maintain social connections later in life. Specifically, loving contact is key to growth of a specific type of nerve cell, the mirror neuron, associated with empathy (more on that later). As psychologist Louis Cozolino of Pepperdine University observes, "Without stimulating interactions, neurons and people wither and die. In neurons, this process is called apoptosis, while in humans, it is called anaclitic depression."

Besides stimulating general brain growth, early interactions with loved ones are crucial to organization of the right brain, a central site for the processing of emotion. The right hemisphere is especially responsive to nonverbal cues, such as facial expression and tone of voice. Developmental psychologists suggest that right-brain-to-right-brain signaling, beginning when infants are around four months old, constitutes the first and most basic language between child and parent. Colwyn Trevarthen, professor of child psychology at the University of Edinburgh, calls these interactions "protoconversations."

These early moments of meeting, if they are positive, tune youngsters' brains to the social channel, teaching them how to communicate their needs and thereby evoke satisfying responses. Babies learn how to get their mother's attention and draw it back when she misses or mistakes their signals—the general process of attunement, misattunement, and reattunement. They learn to attend to the signals their mother sends by focusing on her face and holding still for longer lengths of time; they also learn to read what she wants from them. All in all, we learn in these first interactions whether we can depend on loving responses from another to help us keep our emotional balance. We also get the first glimmerings of how others see us and begin to formulate of our sense of self.

If we are lucky, our first caregiver's expressed delight in us tells us that we are indeed delightful; open responsiveness shows us that we are cared for and valued. If we are unlucky—perhaps our mother is stressed or clinically depressed—we will not get consistent emotional reassurance and nurturing, and we may feel unloved and unworthy. We pick up that we are emotionally on our own. With repetition these exchanges are burned into our brain and form a neural template—a kind of "if this, then that" guide—for our close relationships from childhood into adolescence and adulthood. Positive childhood models tend to enhance our ability to shape adult romantic attachments. And negative models do the opposite. I see it in couples stuck in misery. I ask Marcus, "What happens to you when your wife goes still and quietly tells you, 'I need you so much. I love you'? You bite your lip and turn your head away?" Marcus blinks at me. Slowly he starts to speak. "I am treading water here. There is no place in my brain to put that. If she tears up and tells me she needs something, I freeze. Her tears are indictments. I must have screwed up. If she is upset, she is going to put me out in the cold. Right now, I can't move."

In Marcus's family, his mother's disappointment or tears were always a prelude to angry outbursts, and he remembers as a small child being sent to his room and left there by himself for hours whenever his mother erupted. He learned that another's upset meant that he was flawed and unlovable and about to be deserted. As an adolescent, his solution was simply to stay away from his family and play video games in his room. Now, with his wife, there is no obvious workable solution; everything that he knows how to do makes her more angry and upset.

Some psychologists argue that a person's way of dealing with emotions and relating to others is primarily genetically based—that is, determined by nature, not nurture. A person's innate temperament may incline him or her to be more or less stoic or volatile. But there is growing evidence that repeated patterns of early interaction with caregivers are extremely powerful and can mold lifelong responses to negative emotions and stress. Psychologist Michael Meaney of McGill University in Montreal has discovered that in rats, a mother's intense nurturing of her pups, including grooming and licking, is powerful enough to influence her offspring's effectiveness in regulating fear responses and acting adaptively in the face of danger as adults.

These highly nurtured rats can stay composed even when tightly restrained or stressed, a condition achieved when researchers put them into canisters filled with water to see if they swim or sink! I have an image of the much-loved little rodents lying lazily on their backs with little gin and tonics in their hands, humming a tune called "My Mommy Loved Me; Nothing Bad Will Happen" as they float happily around. Their less-loved cousins, on the other hand, are paddling furiously and screaming their heads off: "How could you do this to me...I will *drown!*" (No, Michael Meaney didn't let the insecure ones drown; psychologists are, in fact, a sentimental lot.) The highly nurtured rats also showed lower levels of stress hormones compared with the rats that received less intense care.

The popular axiom that evolution favors "survival of the fittest" is usually taken to mean survival of the most aggressive. Today, at psychological conferences, we're hearing much more about survival of the "most nurtured." We're discovering that devoted nurturing can overcome the influence of genetic inheritance and even reverse it. And that applies at every step on the evolutionary ladder.

Psychologist Stephen Suomi, who assisted Harry Harlow with his monkey experiments and now heads a major research laboratory at the National Institute of Child Health and Human Development in Bethesda, Maryland, has found that highly reactive monkeys, the genetically primed "bad" boys and girls, actually become skilled leaders and all-around good citizens when cared for by extra-attentive mothers. In general, we now know that no matter what your genetic heritage is, it is repeated experience that turns genes either off or on. The experience of being held and groomed appears to switch off genes that make the brain sensitive to stress hormones and switch on genes that start up calming mechanisms.

This kind of research has recently been extended to humans in studies of so-called "dandelion" and "orchid" children. As described by developmental psychologists Bruce Ellis of the University of Arizona and Thomas Boyce of the University of California, Berkeley, dandelion youngsters have the ability to thrive in all environments. Orchid children, in contrast, are highly sensitive to their environment, especially the quality of parenting they receive. If neglected, they wither; if cared for, they flower with unusual beauty.

In one experiment led by geneticist Danielle Dick of Virginia Commonwealth University, DNA from 400 adolescents who had been followed from birth was analyzed for variations in a specific gene, CHRM2, that is implicated in alcohol dependence, antisocial behavior, and depression. They found that children with the gene variant who had less engaged, more distant parents exhibited the most undesirable behaviors, such as delinquency and physical aggression toward others. But children with the gene variant who had very attentive, involved parents had much better outcomes. They had fewer behavioral problems and a significantly lower risk of depression and anxiety, which are risk factors for future troubles. This points to the power of secure connection to bring out the best in all of us.

Devoted early nurturing grows brains that are better able, years later, to regulate stress, connect with others, collaborate to solve problems, and, of course, dance a meaner tango. The greatest gift a parent has to give a child—and a lover has to give a lover—is emotionally attuned attention and timely responsiveness. The evidence is that throughout life we build on the scaffolding provided by our first relationships to find our emotional balance and link to others. A mother sings in a low voice and softly touches her baby's cheeks as she rocks him to sleep each night; he calms and his heart rate slows. He learns that voice and touch will soothe him, and eventually he can soothe himself just by calling up the memory of the singing and touching. In this way, we gradually develop automatic ways of managing emotion that carry into our adult love relationships. This process also builds expectations about the ways in which emotional moments with our romantic partners will play out.

David, whose mother swung between being high on painkillers and being irritable and abusive, tells me in a therapy session, "I can't get past the feeling of heat here in my chest. I want to escape. It seems like anytime a strong feeling comes up, I want to run. I don't know what you mean by 'comfort.' The only emotion I know about is anger. Maureen's saying she really feels love for me right now as I talk about my fears, but I would never ask anyone to take care of me. I don't know what to do when she gets all syrupy with me. Emotions are private, to be dealt with on your own. What does it matter if she knows how I feel, anyway? What is she going to do about it?" David's response fits with brain research that finds that insecurely attached adults have strong physiological reactions to any uncertainty or to psychological stress, and that avoidants in particular—like David—tend to make many more errors reading their partner's signals, even when they are tender, loving overtures. They have not learned to trust such messages and so cannot use them to calm their fear.

The good news is that we don't have to stay fixed in negative neural pathways. The brain, as we'll discuss later in this chapter, is amazingly plastic, and we can create new neural circuits, altering our ways of perceiving and signaling our emotions to loved ones and revising our expectations of how they will respond to us.

THE NEUROCHEMISTRY OF LOVE

The concept of a love potion, a substance that can turn on love, is found in almost every culture. Many concoctions—made from plants, herbs, insects, animal organs, precious stones, and corals—have been touted. They don't work. But there is one potent formula that our own body manufactures. Called oxytocin (a name that sounds more like a detergent than a font of euphoria), it exists only in mammals and is both a neurotransmitter, meaning it communicates with the brain and nervous system, and a hormone, that is, it communicates with organ systems, too. Oxytocin was discovered back in 1909, but research on this chemical has exploded just in the last decade. The number of Google searches for the word has soared 5,000 percent since 2004.

Scientists have dubbed oxytocin the "cuddle hormone" for its ability to promote strong bonds between mother and infant and between adult lovers. It's also called the "molecule of monogamy" (more on that in Chapter 5). But it's most accurately described as the master chemical of social connection. Both sexes have oxytocin receptors in their brain, but oxytocin levels are generally higher in females. Males have higher levels of a very close cousin, vasopressin (the difference is just two amino acids), which has the same linking effect as oxytocin but also stimulates aggressive behavior, such as mate guarding. We've long known that in humans, oxytocin is released during breastfeeding and orgasm. But with more sensitive assays, we have now discovered that our brain gives us a little dose of the cuddle hormone whenever we are physically near to those we love. In fact, just *thinking* of our loved one will trigger a rush of this hormone.

This chemical packs a punch. A whiff of oxytocin increases our tendency to trust and engage with others in a less defensive, more empathetic way. Anna Buchheim, a clinical psychologist at the

University of Innsbruck, Austria, and her colleagues invited into their lab 26 male students who scored as insecure on attachment questionnaires and gave them a dose of oxytocin. At another time, the students received a placebo. When they received the neurotransmitter, 69 percent of the students responded in a more secure, affiliative way to a series of pictures depicting events such as loss and separation from a loved one. They shifted from agreeing with statements like "I would distract myself and take care of this by myself" to endorsing statements like "I would share with someone and look for support." This shift into secure responses was particularly noticeable in anxiously attached men.

Oxytocin turns off our threat detector, the amygdala, as well as the hypothalamic-pituitary-adrenal (HPA) axis—the "get up for challenge" part of our nervous system—and turns on the calming, "relax, all is fine," parasympathetic nervous system. The effect is to reduce fear and anxiety and lower production of stress hormones. In one experiment, both men and women rated even strangers as more trustworthy and attractive after a dose of oxytocin than they did before taking the neurotransmitter. In another study, 47 couples were given either a placebo or oxytocin before discussing an area of conflict in their relationship. Those who received oxytocin showed significantly reduced levels of the principal stress hormone, cortisol, after the discussion. They also displayed a significantly higher ratio of positive to negative behaviors; partners more frequently held eye contact and agreed with each other and were less belligerent and blaming.

What I see in a couple therapy session is that as one person takes small emotional risks and the other learns to respond, they "re-tune" each other's nervous systems to higher levels of equanimity, which makes them more trusting and flexible. And this is exactly what happens when we first fall in love. "Falling in love" is an accurate phrase. It's a risk, reach, reassurance dance. Oxytocin seems to prompt us to take chances and rewards us with contentment when we find reassurance in our partner's arms. Mutual vulnerability and recovery *with* our lover, accompanied by oxytocin release, is the true tale of love.

All the recent findings support John Bowlby's claim that the bond of love is a safety and survival mechanism and that one of its main roles is to make life less terrifying. And, as with most core survival processes, there is a feedback loop. Oxytocin spawns trust, trust generates closeness and sex, orgasm stimulates oxytocin, and around it goes. Entrepreneurs have been quick to sense opportunity. You can order an oxytocin-infused nasal or body spray called Liquid Trust online. Before you rush to stock up, however, remember that our clever brains tend to adjust for context. Spritzing the hormone likely won't make you feel more loving toward someone you already distrust.

That said, the social effects of oxytocin are still mind-boggling. When dosed with oxytocin, we fixate more on others and gaze longer into their eyes. Scientists suggest that this may be why oxytocin helps us better read others' facial expressions and correctly tune in to their intentions. Let's face it—most signals in love relationships are subtle or ambiguous and require decoding. If you don't think so, try interpreting "I am too tired for sex tonight." We have to decide if that means "I am tired" literally, or "I am tired of you," or "This is the end of your sex life forever."

In a study by psychologist Gregor Domes and his colleagues at Rostock University in Germany, young men between the ages of 21 and 30 were asked to look at pictures of people's eyes after being given at different times oxytocin or a placebo. They were told to pick from a list of words the one that most closely captured the emotional and mental state they saw. After oxytocin, the young men were much more accurate in their readings, even when the expressions were subtle and ambiguous. Domes notes that the hippocampus, the region of the brain that is key in retrieving memories, is very rich in oxytocin receptors. He suggests that oxytocin may help pull up stored images of expressions that aid people in interpreting what they see at a given moment. The adaptive advantage of such a chemical boost is obvious when it comes to romance. The accurate interpretation of nonverbal signals allows us to tune in to and effectively coordinate moves with our partner to create a harmonious dance.

As if this isn't enough, oxytocin receptors are also plentiful in the part of the brain—the nucleus accumbens—that is central to production of dopamine, the neurotransmitter that makes us feel elated and euphoric. Researchers believe that oxytocin increases release of dopamine, further supporting attachment between partners. We tend to stick around people with whom we feel pleasure.

Because dopamine activates the same neural circuitry as cocaine and heroin, some scientists have wondered if love could be viewed as an addiction. There are similarities: when we are smitten, contact with the beloved produces positive feelings, just as a drug does; there is a "hunger" for contact with the loved one, and there is distress when this hunger is not satisfied. But an addiction is a negative, costly, compulsive behavior that constricts a person's life and behavior. Positive romantic love, on the other hand, expands our world; it makes us more confident, flexible, and open. Moreover, secure connection seems to protect *against* addiction. Recent research at Duke University shows that rat pups who are touched frequently by their mother have higher brain levels of interleukin-10, a molecule that suppresses a craving for morphine. Similarly, bonding in monogamous prairie voles appears to decrease the rewarding effect of amphetamines in the brain. I have an image of a little rodent wiggling his whiskers and whispering to his spouse, "I don't need no drug but you, my voley baby."

THE NEURONS OF LOVE

While some researchers are focusing on brain chemicals, others are concentrating on different types of brain cells and delineating their roles in love relationships. Here, too, the findings underscore the fact that nature has wired us for connection. We are much more of a social animal than our individualistic society has recognized. As a master researcher in this area, Marco Iacoboni of the David Geffen School of Medicine at UCLA, asserts, "We are not alone, but are biologically wired and evolutionarily designed to be deeply connected to one another."

One of the most exciting areas of current inquiry centers on mirror neurons, so named after a remarkable and serendipitous discovery in the 1990s in the laboratory of neurophysiologist Giacomo Rizzolatti at the University of Parma, Italy. A member of his team—Vittorio Gallese or Leo Fogassi; no one is sure at this point—was moving around the lab while a female macaque with electrodes in her brain sat quietly in a chair awaiting her next task in an experiment on motor control. The researcher idly picked up something—a peanut or ice cream; no one is certain which one. Suddenly, a burst of sound came from the computer that was recording the monkey's brain activity. Even though she wasn't doing anything, her brain had lit up as if she were the one picking up the food!

Researchers had accidentally stumbled on the solution to a mystery that philosophers have struggled with through the ages: How do we know what is happening in the mind of another? The answer: mirror neurons. They put us inside the body of others, making us literally feel what they are feeling. Mirror neurons explain why we shrink back in our seats with fear when the hero is abruptly attacked by Freddy Krueger in *A Nightmare on Elm Street* and why we soar with joy when the young bicyclists lift into the blue sky in *E.T. the Extra-Terrestrial.* These neurons are kicking in when we wince after a kid tumbles off a swing onto the ground and when we break into a smile watching a friend's eyes light up as we carry in a huge birthday cake.

This ability to enter into another's experience is especially pertinent in love relationships, in which responding in a sensitive way to a partner's needs is so central. When we see our sweetie's mouth droop down or eyes well with tears, our brain mimics the experience for us. In a sense, we physiologically try on the feeling. The line between us and our partner blurs, and we automatically, without conscious reflection or deliberation, feel and know he or she is sad. This is invaluable in helping us tune in to a mate and in building intimacy, safety, and trust—the very bonds of love.

This exquisite sensitivity begins when we are about two years old, at about the same time we start to be able to recognize ourselves in a mirror. "Knowing me" and "knowing you" are linked; they are two sides of the same coin. So how do we distinguish between our feelings and those of others? Our mirror neurons take care of this for us as well. A subset of these brain cells, super mirror neurons, fire more rapidly for our own experiences and more slowly for the experiences of others. The brain is a perfectly honed social device, supporting our sense of self while seamlessly linking us to others.

Mirror neurons do more than just mirror the observed actions of others. They clue us in to their intentions. Mirror neurons don't fire if we see someone aimlessly pantomiming or pretending with no real purpose in mind. For example, if someone acts as if they're reaching for a pen, but there is no actual pen, our mirror neurons remain dormant. But they will fire when they sense a goal, even if it is not completed or is slightly modified. In a sense, they fill in the blanks; they predict and anticipate. Say someone is going to close his hand to pick up a pen, but he is distracted and pauses; then he opens his hand to pick up pliers instead. Mirror neurons will have fired as the person reaches, because the intention—to pick up an object—remains the same. Mirror neurons are our "intention radar," which allows for the instant coordination of complementary responses with another person.

In love relationships, mirror neurons are how we automatically "know" what our loved one will do. Marie's distress has registered with her husband, Simon, and his face is now sad and concerned. She sees that, and her facial muscles duplicate what she sees. His hand reaches out, and she knows he is going to stroke her arm. She bends toward him. He laughs and says, "Maybe I was just reaching for my wineglass."

"No," she responds. "You were reaching for me." Simon smiles and pulls her into his arms, and they hold each other. A small moment, but so much has happened. And so easily.

Mirror neurons have upended our assumptions about how we read each other. We used to think that Marie would stay in her head, reasoning out what Simon would do. But now we know that we comprehend each other's intentions in less judicious fashion. In a flash, Simon felt what Marie felt, and Marie felt what Simon felt—and she got his intention. Such moments of connection are the lifeblood of love relationships.

Often when I am training therapists, I do a demonstration session with a couple I have just met. Members of the audience are mystified that I can tune in to each partner's deeper emotions so fast and so easily. They ask me how I know what a partner is feeling. Do I have an algorithm of feelings, an "if this, then that" list? I do, but usually it is not up and running in my brain. There is no effort involved. If I stay calm and am attentive to the partners' gestures, tone of voice, and facial expressions, I can feel what they feel even when they cannot give it a name. I can see fear and the intention to turn away before one of them even says, "I don't think it's useful to talk about this."

EMPATHY IS US

The torrent of research into the brain is presenting us with a dramatically different view not only of love relationships but also of human nature. Western society has long held a rather pessimistic view: we are essentially insular, selfish creatures who need rules and constraints to force us to be considerate of others. Today, we are drawing a diametrically opposed portrait: we humans are biologically driven to be associative, altruistic beings who are responsive to others' needs. We should, it seems, be called *Homo empathicus*.

Empathy is the capacity to perceive and identify with another's emotional state. The word, coined in the 20th century, derives from the Greek *empatheia*, meaning "affection" and "suffering." But the concept was first developed by 19th-century German philosophers who gave it the name *Einfühlung*, meaning "feeling into." Empathetic concern in higher mammals probably evolved from a need for flexible, adaptive parenting to assure survival of the young and from a need to collaborate in defense and hunting to assure a pack's or tribe's continued existence. How strong that capacity is in human beings is being proven in study after study.

Most fascinating, perhaps, is research showing that just imagining or thinking that another person, especially a loved one, is in pain makes us respond as if we are going through the exact same experience. In one experiment, neuroscientist Tania Singer and her colleagues at the University of Zurich found that when a woman received a small electric shock to the back of her hand, the woman beside her, who received no shock, reacted as though she had received it, too. The identical area of the brain lit up in both women; the same pain circuit was activated. We literally hurt for others.

Roughly, the way empathy seems to happen is: *you see me*—or even, as in the experiment above, *imagine me*—experiencing a strong feeling, maybe pain or disgust; *you mirror my response* in your brain; *you mimic me* with your body (your face crinkles in the exact same way as mine does); *you respond to me* on an emotional level and move into empathetic concern for me; *you help me*. As we imitate others, we also communicate and show them that we feel for them; this creates instant connection. In Oregon State University psychologist Frank Bernieri's study of young couples teaching each other made-up words, pairs who showed the greatest motor synchrony, that is, those who mimicked each other most closely, also had the strongest emotional rapport with each other. In my own team's studies of forgiveness, nearly every injured partner told his or her lover some version of, "I can't forgive you until I see that you feel my pain. Until I know that my pain hurts you, too."

Empathy is not limited to humans. Primatologist Frans de Waal, in his book *The Age of Empathy*, lays out a clear case that all species who have mirror neurons and a sense of self (that is, they can recognize themselves in a mirror) including humans, dolphins, apes, and elephants—respond to each other's pain and grieve when one of their number dies. In other words, they show all the signs of emotional bonding and empathy. Rhesus monkeys, for example, refuse to pull a chain that gives them food if this act delivers an electric shock to their neighbor in the next cage. The monkeys starve themselves to avoid inflicting pain on another. Elephants walk miles to mourn at the grave of a herd member, and chimpanzees offer solace, hugging, holding, grooming, and calling to a stricken relative who has lost a fight with an older member of the troop.

It is distressing to see others in pain, especially if they're familiar to us. So why do we sometimes have such trouble feeling empathy for those we love? Three possibilities, gleaned from the everyday experiment called a couple therapy session, present themselves. First, a person's mirror neurons may be underdeveloped or functioning poorly; failure of the mirror neuron system has been associated with an inability to resonate emotionally with others, as in autism. Second, stress or depression may have exhausted a person's mental resources, so that he or she is essentially emotionally numb. Depression and stress hormones, such as cortisol, have been shown to impede brain growth and even damage its social and emotional centers. Abuse early in life tends to shrink the hippocampus, the area of the brain that deals with ordering experience into coherent emotional memories. As a result, the brain becomes more sensitive to emotional stressors, such as separation anxiety, but has a less developed neural network for containing such anxiety.

The third and most common reason is that we simply become distracted. Some preoccupying emotion, such as the overwhelming fear of upsetting or losing a partner, blocks the ability to focus on the other's anguish. It is hard to concentrate on another person when you are spending all your brainpower just trying to calm yourself. As I was tumbling through a terrible storm at thirty thousand feet in the small silver tube called an airplane, all my empathy training deserted me when the large man sitting next to me suddenly said, "I think I am going to have a panic attack." I heard myself say, "No. You're not. Just stop it." I only had time for a little guilt when we were safely on the tarmac.

The bottom line is that how well you are able to deal with your own emotions will greatly affect your ability to tune in to and feel empathy for others. The attainment of emotional balance allows us to engage with, feel for, and respond to the concerns of others. Being securely attached furthers this balance. Psychologist Omri Gillath of the University of Kansas looked at what happens in the brains of women with different attachment styles when they are dealing with difficult emotions. He asked them to first imagine emotionally neutral events, such as shopping with their partner; then he asked them to imagine everyday arguments; and finally he asked them to call up painful relationship scenarios, such as their partner leaving them for someone else or dying. Then he instructed them to stop thinking about these things.

He found that the more anxiously attached women had more active emotional brains than the other groups did when thinking about the painful scenario. In particular, the anterior temporal pole, which calculates the emotional significance of events and, especially, processes sadness, really got busy. Meanwhile, the orbital frontal cortex, which regulates emotion, took a nap. Without a braking mechanism, the women's brains simply continued ruminating on the painful event. Women who were avoidantly attached weren't able to suppress their thoughts and feelings very well, either. Key emotional areas of their brains stayed active. This parallels previous research by Mario Mikulincer, who found that avoidant people's ability to suppress emotion is incomplete and easily disrupted by any kind of mental task—for example, being asked to remember a seven-digit number. Under even this minimal stress, thoughts and emotions rebounded, and emotional balance was compromised.

Gillath's brain-scan study affirms numerous other studies showing that more securely attached people fare best in dealing with difficult emotions. They can generally tolerate them better and regulate them more effectively; they are less likely to be swamped by or obsessed with their feelings or to labor excessively to deny them. Being flooded with negative emotions or constantly working to numb ourselves inevitably distracts us from being emotionally present with others, from tuning in to others' feelings and needs. We have to have some measure of calm and security within ourselves before we can be sensitive to and caringly respond to others.

Pete, who has PTSD as a result of childhood abuse, has known only one safe relationship—the one he has with his wife, Sally. He is so terrified of losing her that when she is upset with him over even the most minor thing, he blanks out emotionally. He cannot pick up on Sally's cues. "This stuff is like a foreign language to me," he tells me. "I don't get it. It doesn't compute." After a few therapy sessions, during which we worked on getting him to stay calm while he focuses on Sally's face, he seems better able to resonate with her emotions. He looks sad as she cries and squirms in his chair as she complains, but he is constantly sidetracked by his own strong physical sensations of panic. I ask him, "What do you see in her face?" He replies, "All I know is that she is angry with me, and all I want to do is stop it, fix it. I am thinking of all the ways out, and then it just becomes too much, so I freeze up." He cannot tune in to or name—much less tolerate—his own emotions or Sally's.

Perhaps his brain, like those of institutionalized Romanian orphans, has not fully developed the neural pathways that allow him to stay present and be empathetic to his wife. Perhaps it is simply fear, which focuses us in on ourselves, blocks sustained attention to others, and prevents the decoding of their signals. Perhaps someone like Pete, who never experienced secure attachment as a child, is extremely sensitized to threat and has only inefficient ways to calm his attachment fears. This neural pathway of fear and loss is the deepest, most traveled, and most easily accessed in Pete's brain. As he falls into attachment panic and tries to deal with this emotion by shutting down, he automatically draws negative responses from his partner. Pete's mirror neurons are not in top form, so he has no sense of his impact on Sally or how he sets up interactions that confirm his worst fears.

However, when a dance goes wrong, it is almost never due to just one person's brain patterns, emotional style, habits, or expectations. The other person's responsiveness or lack of it always plays a part. Amanda Vicary and Chris Fraley, psychologists at the University of Illinois, asked individuals to imagine themselves in a relationship story and, at twenty predetermined points, choose between options that describe what they would do next. For example, at one point, after the subjects are told that their partner has been talking on the phone with a former lover, they can select either "I am glad to know you still get along with people you have dated" or "Is there still something going on between you two?" These choices, in essence, forced the participants to interpret their supposed partner's intentions.

In the first phase of the study, the story was written so that the selection of a given option—positive or negative—didn't affect how the imaginary partner responded. The imaginary partner was impervious and unresponsive. The insecure folks then consistently kept choosing statements that showed they distrusted their lover. In the second phase, the researchers shaped the story so that the partner was occasionally supportive. That show of concern was enough to slowly change insecure folks' attitude toward their hypothetical partner. They began to judge the partner's intentions more favorably and to select the more affirmative statements at the predetermined points. Interacting with a warm partner made everyone, secure or insecure, more positive. Just as the potential to maintain negative patterns exists, so does the potential for change.

THE PLASTIC BRAIN

One way to change our behavior is to change the brain. Scientists used to firmly believe that we were born with a finite number of brain cells and that when any were destroyed, we were out of luck. But the brain, it turns out, is much more plastic. We grow new neurons and new links between neurons throughout our lives as we have new experiences. So can we grow Pete's brain in couple therapy by shaping new experiences for him? Perhaps. Every experience changes our brain somewhat. One way to create new neural pathways is to block our usual ways of behaving. For example, if we were to wear a blindfold, within two days our visual cortex would begin to process signals from touch and hearing.

So maybe we can help Pete replace his "freeze" reaction with a "tell Sally what the fear feels like" response. Once he is better at modulating his own fear, we can then help him to see and respond more caringly to Sally's pain. We can help him focus by asking him, "What do you see right now on Sally's face? Can you feel in your body what this feels like for her?" And we can help Sally to send clearer emotional signals to Pete. We can also increase the elements that promote empathy—perceived familiarity and similarity—by getting partners to recognize their mutual longing and vulnerability. Finally, we can help them express the empathy they feel. We can see we have made headway when, in our twelfth session, Pete says to Sally, "I am getting now that we are both scared, that you are vulnerable, too. I see it in your eyes and see that it hurts. I don't want you to feel scared. I don't know what to do, but I want to comfort you. "

Our neurons are always ready to sing the song of connection with another. It seems that all through adulthood there is the possibility of the flowering of new neurons and networks in our brain as we have experiences of love and loving. Walter Freeman, professor of neuroscience at the University of California, Berkeley, agrees. He concludes from his survey of learning across the human life span that there are two major events that naturally create massive neural reorganization: one is falling in love and bonding with a partner and the other is beginning to parent. (Any parent will tell you that the first few months of taking care of baby definitely rearrange your brain!) Freeman suggests that oxytocin is a key "neuromodulator" that can enhance or diminish the overall effectiveness of the connections between nerve cells.

Since the brain is fundamentally an organ of socialization, it makes sense that there are heightened periods of plasticity when the brain is rapidly adapting to new circumstances. New connections with a caring other shape and reshape us; new neural connections, new emotions, new understanding, and new moves in the dance of love can offer us a new world and sense of who we are. Pete tells Sally in our last session, "Now that I know how to reach you, I feel different. Bigger somehow. Less careful. This new way we have of being together is changing me. Seems like I can explore more now, take chances."

I think of passages in Louis Cozolino's book *The Neuroscience of Human Relationships*. Cozolino, a psychologist at Pepperdine University, points out that early relationships optimally sculpt key parts of our brains, such as the prefrontal cortex, in ways that allow us to "think well of ourselves, trust others, regulate our emotions, maintain positive expectations, and utilize our intellectual and emotional intelligence in problem solving." Now we know that it is not just early relationships that have this effect.

We now understand key aspects of the chemistry of love, such as the function of the cuddle hormone, oxytocin, and the workings of the neural pathways of human connection. We are beginning to see that part of the "fire" of love is the firing of key nerve cells—the mirror neurons, which connect us with others. But more than this, we now understand that not only does biology shape love relationships, but these relationships in turn also shape and regulate our physiology. A kiss can bring a cascade of oxytocin and dopamine that shuts off stress hormones, brings down our heart rate, and tunes up our brain's ability to read our lover's face.

This kind of science goes a long way toward explaining why the brain itself is now considered a social organ. Robin Dunbar, professor of evolutionary psychology at the University of Oxford, observes that established explanations for why we evolved with such huge brains are inadequate. Such brains are exceedingly expensive in terms of the energy needed to keep them going, and most animals survive just fine with brains that are much smaller. Dunbar suggests that our brain size is not so much the result of having to master technical problems such as creating tools or shelters as it is the consequence of our having to engage with others to survive and thrive. Social interaction is a complex chess game: we have to anticipate the actions of others as well as the effects of our own actions on them, and we need lots of brainpower to do that. Thus our brains grew larger and ever more designed for connection.

Shakespeare asks, "Tell me where is fancy bred . . . in the heart or in the head?" Scientists reply firmly, "The head." But some of us would put the main source of love much lower. "Love is the poetry of the senses," rhapsodized Honoré de Balzac. American writer Harlan Ellison puts it more bluntly: "Love ain't nothing but sex misspelled." That's one hell of a misspelling, but let's examine this issue in the next chapter.

EXPERIMENT 1

Set aside thirty minutes when you and your loved one can be together quietly. Choose a time when you are both feeling relaxed—that is, not just after a fight. And choose a place where you will not be interrupted.

Begin trying to tune in to each other. Stand face-to-face and almost toe-to-toe, but not touching, and look at each other's chest. Then synchronize your breathing. Find a steady rhythm.

Then one of you should speed up the rhythm, and the other should try to match it. Once you are in sync, reverse roles. But this time, the "leader" slows down the breathing rhythm.

Continue the breathing exercise through three more exchanges.

Then add a new element—shifting weight from one foot to the other. Start out breathing in unison and moving onto the same foot as the other person.

Once this is fluid and easy, the taller person should begin to gently turn his or her shoulders to the right or left and then back to center. The other person follows, matching the pace and movement.

After a while, switch the lead.

Once you are again in tune, begin turning shoulders so far that you both begin to lose your balance. The "follower's" job is to repair the loss of balance with a little pressure from a hand placed on the leader's arm. Do this five times, then switch the lead and do it five more times.

Sit and talk about whether moving into physical synchrony made you feel more emotionally connected to your partner. If it did, describe how that feels so the other person can understand. If it did not, see if you can pinpoint what got in the way of getting to a more tuned-in place.

EXPERIMENT 2

This experiment explores your level of empathy and your ability to read your partner's intentions.

A. Imagine you are wanting to talk with your partner about the job interview you had a few hours ago. You are worried about it. But your partner seems distracted. You begin to talk about what happened, and he or she begins to give you advice about the things you might have done at the beginning of the interview to make a good impression.

Do you say:

1. "Maybe you don't want to talk with me about this. Why don't we just leave it for now?"
2. "I am really worried about how I did. I just need some comfort and reassurance."
3. "You are telling me all the things I did wrong and you are not even listening. You don't listen."

Which of these three possible responses is the most relationship enhancing and likely to elicit empathy and a loving response from your partner?

Now imagine that whatever your reply, your partner says: "Oh, I'm sorry. Maybe I am not really tuning in here. You probably want me to just support you. Those interviews are hard."

What do you say next?

Even though your partner first disappointed you, did you subsequently respond positively to the later offer of care? Discuss with your partner how easy or hard it was for you to accept this attempt to repair the disconnection between the two of you.

B. Your partner admits forgetting to arrange for a tradesman to visit next week, as you had asked. Your partner sighs and murmurs, "Today just felt like everything was going wrong in my life. I felt down."

You say:

1. "That really worries me. I rely on you to do the things you say you will do. But you do look kind of down. Maybe you should start going to the gym. You'd feel better."
2. "The tradesmen need time to make an appointment. I think you should call soon."
3. "You look upset. Everything went wrong? Do you want to talk about it? You look really down right now."

If you couldn't manage the most empathetic response—you know which one it is!—reflect on what made that response difficult for you.

Let's presume that your mirror neurons fired. Did you not connect emotionally with the information your neurons sent? Were you caught up in and distracted by your own feelings and unable to focus on your partner's cues? Did a belief or judgment, such as "You do your tasks no matter how you feel," get in the way? Do you feel anxious or unsure of what to say when your partner confides a vulnerable feeling?

No one can be empathetic all the time, but if we want to build a loving bond, we do need to know what blocks our empathy and learn how to tune in and respond.

CHAPTER 5

The Body

Sex is emotion in motion.

—Mae West

ho would believe that Mae West, the bawdy enter-
tainer known for her racy quips and double entendres,
would hit on the essential truth of sex? It *is* emotion—the quality of our connection to another person—that
defines the type of sex we have, the satisfaction we derive from it,
and the impact it has on our romantic relationships. Indeed, attachment determines how we behave in bed as well as out of it.

This is a radical idea. For years, sex has been central in our
beliefs about adult love. Freud started this conviction with his
theory that the physical pleasure we gain from our opposite-sex
parent's nurturing and cuddling in childhood is an erotic bond
that becomes the template for our adult romantic relationships.

Later researchers—such as Alfred Kinsey and William Masters and Virginia Johnson, with their inquiries into sexual experience, mechanics, and biology—pushed sex into further significance. The women's-liberation movement unwittingly endorsed this view with its proclamation that women are entitled to have as much sex and draw as much pleasure from it as men—and the Pill, which freed them from fears of pregnancy, allowed them to do so. In recent years, evolutionary biologists and psychologists have made sex even more prominent with their theory that love is simply a trick—nature's way to induce us to have sex and thereby assure continuation of our species.

As a result, in the Western world we've come to believe that sexual infatuation and love are synonymous and that sex is the essence of adult love. In simple terms, sex is love—and good sex is good love. Today we are obsessed with how to have good and ever-better sex. The best sex, of course, is orgasmic. If you doubt the current state of affairs, pick up any women's or men's magazine and you'll find at least one, and likely more than one, article detailing techniques and positions to liven up your sex life. Visit any bookstore and you'll find tome after tome offering the "secret" to firing off the Big O, which, preferably, looks like the kind of seizure you might have if you put your finger in a light socket. Companies, meanwhile, have introduced a spate of new products, from ribbed and flavored condoms to spiced-up lubricants to toys and aids guaranteed to whip you into sexual frenzy and satisfaction. Sex seems to be portrayed as a process similar to digestion, as psychologist Leonore Tiefer of New York University School of Medicine suggests, rather than what it is—a reciprocal dance.

Vaulting sex to such primacy has, alas, distorted its role in relationships—and with harmful consequences. Instead of drawing people closer together, all the emphasis on sex is instead driving us farther and farther apart. Consider the fixation on Internet porn. We're abandoning living partners for screen sex. Forty million Americans admit to being regular visitors to Web porn sites, and 10 percent of them say they are addicted. Although patrons of these sites are mostly men, women are fast catching up. And what's most troubling is that the followers are getting younger and younger—teens and even preteens are watching Internet porn. When adults do get together today, it's often for one-night stands or casual sex. They call themselves friends with benefits. They're going through the motions but with little emotion.

The sad fact is that we have isolated sex, taken it out of context. Yes, it is an important aspect of romantic relationships, but it is not the be-all and end-all. To researchers like me, adult love has three elements: sexuality, caregiving (a blend of attentiveness and empathy), and attachment. And the last is by far the most important. *For as we connect emotionally, so we connect sexually.*

The level of ease with closeness and the degree of safety we feel with our partner translates into different kinds of sex, each with its own practices and goals, and it even directs our sexual fantasies. We can have sex that centers on physical sensation and is walled off from our heart, our emotional life. We can have sex that is mainly emotional consolation, focused on comfort and relieving our fears. Or we can have sex that is synchronous, intimate, and integrated with our deepest emotional needs.

This idea was brought home to me recently, albeit in an altogether different context. It was Friday night at the *milonga,* a social gathering for people who dance the Argentine tango. As the strains of the *bandoneón* and violin filled the hall, men and women took to the floor and stepped and swayed. My feet hurt, so I was sitting out and watching. I found myself thinking of my friends' comments that tango is a very sexy dance. They're right, of course. But what is it that makes it so sexy? Is it the close embrace with partners' heads and torsos pressed together, the caressing and entangling of legs and feet, the stiletto heels that make women's legs seem to stretch forever?

Yes, it's all of that. But still, not all dancers give us an erotic charge. Why is that? What makes one pair mesmerizing and arousing and another not? I thought up a little experiment. As my fellow dancers—amateur *tangueros* and *tangueras* all—flowed past me, I rated them on a "torrid" scale. I closed my eyes and opened them at random and scored the couple I had in front of me.

The first couple I saw was new in town. The two were slim and spiffily dressed, he in a fitted suit and two-tone shoes, she in a slinky red dress and strappy four-inch-high black suede sandals. They danced with technique and elegance. He swiveled his hips and arced around her in a masterful *molinete;* she pivoted on her left foot and kicked her right foot high in the air in a stunning *boleo.* Their performance was impressive. But it was just that—a performance. They were dancing for their audience, not for each other, and while I admired their skill, I was left unmoved.

The next time I opened my eyes, a young, casually dressed couple was in view. The dancers walked to the beat, changed weight, and moved into some complicated steps, but there was an awkwardness about them. The young woman seemed nervous to me. She was trying very hard to follow her partner's lead. When he stretched out his foot, she hesitated a beat before stepping over, then looked up as if to ask, "Okay?" I watched, but there was no heat. They were doing steps, trying very hard to do what the other person expected, but not truly dancing.

The third time I opened my eyes, I saw one of my friends, in her usual plain dress and flat practice shoes, dancing with a short, chubby guy in a gray T-shirt and blue jeans. They were doing slow, simple steps, but they were mesmerizing. I couldn't take my eyes off them. It was like they were having an intimate conversation. He turned his shoulders, inviting her into the space he'd made; she accepted, and they twisted into a tight turn together. The music slowed, and he waited considerately for her to finish pivoting and step over his foot and into his embrace. He caught her foot with his and slid it behind her; then she caught his and moved it back. They were playing! They were completely absorbed in each other and in savoring each step and moment of their dance. Their moves were simple, synchronized, and wholly sensuous.

SEX FOLLOWS CONNECTION

Our culture endorses the idea that sex brings emotional attachment, that it creates the bond that ties a couple together. In short, love follows sex. But much more significant is the movement in the other direction. Numerous studies over the past ten years show how the three attachment styles—secure, anxious, and avoidant—influence our motives for having sex, our sexual performance and satisfaction, and the impact of sex on our love relationships.

Those of us who are avoidant, that is, uncomfortable with emotional closeness and dependence on others, are more likely to have what I term "sealed-off sex." The focus here is on one's own sensations. Sex is self-centered and self-affirming, a performance aimed at achieving climax and confirming one's own sexual skill. Technique is prized; openness and vulnerability shunned. There is little foreplay, such as kissing or tender touching. And no cuddling afterward—once the Big Bang occurs, there's nothing left. Partners' feelings are deemed insignificant and are easily dismissed.

Because pleasure without emotional engagement is shallow and fleeting, this kind of sex needs continual boosting to be thrilling. Novel techniques and new partners can momentarily heighten excitement, but the incessant experimenting can lead to unsafe practices and coercive pressure being applied to partners who are hesitant to participate.

Sealed-off sex is most common among heterosexual men (the quintessential practitioner is James Bond) and also can be frequent among gay men, especially if they are not out about their orientation. Disturbingly, this type of sex may be increasing because of the wide spread of Internet porn. Youngsters who troll sex websites are learning about performance and sensation but not about emotional connection.

Henry, a 40-year-old fitness instructor, tells me, "Sex and love are separate—and anyway, romantic love is a con, a fiction." When I ask about his sex life, he says he masturbates frequently. "It's easier than dealing with my wife. And I get frustrated because she never wants to do blow jobs. She knows that they are my big turn-on, but she just won't do it." He goes into great detail about the moves that arouse and satisfy him. He adds, "I guess she is also angry about the one-night stand I had while I was on a business trip. I don't really see what all the fuss is about. Everybody does it. It was just sex."

Alison, an elegant career woman in her early fifties, tells her husband, Michael, "I am tired at night, and if we make love, I just don't want this big demand for cuddling and kissing afterward. I don't enjoy it that much, frankly. I just want to have the orgasm and then go to sleep." But then she turns to me and says, "We also fight less if I have sex with him. We have fewer of these long, drawn-out relationship discussions where he wants to hash over everything. So sex works. But then, you know, it doesn't really impact things. Next day we just go on as usual. The relationship doesn't improve, really."

Sealed-off sex is one-dimensional and leaves both partners dissociated. It undermines emotional bonds. It is also, in the end, less satisfying. Research indicates that it actually reduces arousal and results in less frequent orgasms. In a dance without connection or the ability to tune in to the emotional music, boredom and emptiness follow every step.

More anxiously attached people, by contrast, tend to have "solace sex," that is, to use sex as proof of how much they are loved. There is emotional engagement, but the chief feeling is anxiety. For such people, who are highly vigilant and sensitive to even a hint of rejection, sex serves as reassurance that they are valued and desired. For men, it is usually the sex act itself that gives comfort. For women, it is the kissing and cuddling that precedes and follows it.

Leon, 55, a high-powered lawyer, wants to make love to his wife, Jolene, every morning and evening. He explains that he is highly sexed because his testosterone level is especially high. He adds, though, when he is alone with me, that he is always scared that his wife does not really love or desire him. Even though there is no evidence for it, he obsesses over the idea that she might have had an affair sometime during their thirty-year marriage. He says, "Jo withdraws from me. If only she would want me more, then the relationship would be just fine. Then I could 'rest.'"

When Jolene reminds him of the great sex they had last week, he agrees, but immediately fixates on the week before, when she had turned him down. "I know on some level that I am pushing her and getting kind of demanding," he says. "But I just want to be closer, more sure of her. When we make love, then it's like the sun comes out and I start to feel truly loved. But when she's tired and doesn't want to, I can't help it—I take it really personally, and all my fears whip up." Like other anxiously attached folks, Leon is so sensitive to any relationship threat that he tends to leap to catastrophic conclusions at the first sign of disappointment, sexual or otherwise.

Claire, 38, a petite high school teacher, confesses that she never says no when her partner, Terry, wants sex. "I just try hard to please him. But I guess I have always been of two minds about sex, really. It's hard for me to just relax and let go. I like the closeness, though. The holding each other. The romance is important. I know he loves me then. I am never quite sure why he finds me attractive, you know? I don't see myself that way. He asks me what I want in bed. But I don't really know. I want what pleases him. And I worry that I am not sexy enough for him." Claire is thinking about having an eye lift and liposuction to enhance her looks and desirability. Seeking cosmetic improvements is fairly common among anxiously attached women.

It makes perfect sense that our basic comfort with closeness and vulnerability affects how we express and experience sex. We are wired to put safety first. If we have to constantly monitor our partner's level of love for us, we are distracted from the attunement and responsiveness that good sex requires. We can't be flexible and coordinate our response; we lose the ability to lead and to follow.

Attachment goals and perspectives haunt private sexual fantasies. We see this very clearly when people are insecurely attached. Psychologist Gurit Birnbaum and her colleagues at Bar-Ilan University in Israel asked 48 couples to fill in attachment questionnaires and to keep 21-day diaries of their sexual thoughts and fantasies. More anxious men and women fantasized about their partner being very affectionate in sex, reflecting the yearning for love and reassurance that pervades their sexuality. Mary, who is extremely anxious, described her fantasy this way: "I am with my boyfriend on a secluded beach, and he tells me how much he loves me, caresses me gently, and I feel I have melted in his arms. I hope it will never end." Partners who were more avoidant imagined themselves or others acting in alienated and aggressive ways. David wrote in his diary, "I am at a private party with three amazing naked women, and I'm giving them the time of their lives."

On days when partners reported conflict or criticism in the relationship, anxiously attached partners portrayed themselves as humiliated and helpless. In his imagination, Carl became literally powerless: "She undressed me and tied me up with ropes to the bed, leaving me completely helpless. She was totally in charge and made me a slave of her desires." Avoidant partners, however, depicted themselves as aloof, remote, and invulnerable to the dangers posed by others. Morris described a fantasy in which "some beautiful woman takes my pants off under a table in a library...the librarian stares at me very harshly, but then joins in. I am sure it's the best sex they have ever had. But then they freeze in fear. My girlfriend is watching, so I guess the party is over." Here, Morris is distant and focuses on his superior performance in sex. The result of being caught in an infidelity is simply that the party ends.

Of course, attachment style also influences how sex, whether good or bad, affects our relationship. This is particularly important, since how we make sense of the inevitable sexual failures and disappointments that we all experience will partly define our overall relationship with our lover. In Birnbaum's study, researchers asked how relationship behaviors and satisfaction were affected by having had sex the previous night.

Anxious attachment seemed to amplify the effect of both good and bad sex. This fits with what Kate tells me in my office: "Even if really good sex is not all that frequent, when we do have it, it restores my confidence in us for a while, and I can believe that he loves me. I am more affectionate then. But if the sex seems just ho-hum, or he isn't really turned on—you know, if we are both tired or something—I become really bothered about it. I think about it all the next day, and I get edgy. I tend to push him for attention, and all my worries come up about him not really loving me. It usually ends in a fight."

Avoidant attachment, on the other hand, appeared to dampen the effect of physical intimacy. Sex and the response to one's partner the next day were unconnected. I see this in my practice. In my office, Tom chastises his wife, Anabelle: "I don't know why you are making this big deal about our lovemaking on the weekend. Yes, it was good. So what? Does that mean I am supposed to go around hugging you for days on end? Sometimes it's not so good, so I just forget about it. Sex is just sex." Tom dismisses the attachment significance of sex, alienating his wife in the process.

Sexual satisfaction for both the anxiously attached and the avoidant is constricted: the anxious partner is preoccupied with being loved, and the avoidant partner is determined to stay detached. Worry and distraction do not make for expansive, fulfilling sex. Sealed-off sex tends to be erotic but empty, while solace sex is soothing but unerotic. The most satisfying and orgasmic sex, what I call "synchrony sex," occurs when partners are securely attached.

A secure bond is characterized by emotional openness and responsiveness in the bedroom as well as out. That leads to better communication and engaged, focused attention, which in turn leads to greater arousal, pleasure, and satisfaction. This is sex at its most rewarding.

Secure partners are able to express their needs and preferences. But you'd never know from the images on TV and movie screens that communication is part of sex. There, good sex is almost always a dreamlike experience. Partners never seem to talk; they appear to instinctively know what to do. In the real world, great sex is often full of chatter and laughter ("Move over—I'm falling off the bed").

Many studies now attest to the fact that because secure partners feel safely connected to their lovers, they can access the full richness of their sexuality. Feeling protected gives them the freedom to explore and be sexually adventurous. Think about it. If you trust that your partner is there for you, then you can relax and let go without fear of embarrassment or rejection. Safety fosters a willingness to experiment, take risks, and be fully immersed in the sexual encounter. Sex becomes more spontaneous, passionate, and joyful.

At the end of repairing their relationship in couple therapy,

Elizabeth comments, "You know, I can't believe the difference this work has made to our sex life. I didn't even hope for that. I am such an inhibited person, but now I feel so sure of Peter I can't believe the risks I am taking. Last week I actually asked him for oral sex!" She giggles, then continues, "He didn't seem to mind at all. For me, being able to do that opens up a whole new world. Maybe I am more passionate than I thought I was!"

Unlike insecure partners, secure lovers tend to be confident about their physical attractiveness, sexual desirability, and skill. Studies indicate that the more secure you are, the more you believe that you can control the quality of your sexual experience—that it is up to you rather than your partner or factors like where and when you have sex. This kind of efficacy is empowering and translates into a more active and flexible response.

As with other aspects of a relationship, attachment science gives us a clear picture of exactly what a healthy, optimal sexual relationship looks like. And that offers us a compass. As Yogi Berra said, "If you don't know where you're going, you will wind up somewhere else." Securely attached people report that they prefer to have sex in a committed relationship and that affection and expressions of love are key parts of their sexual experience. They report more passion, pleasure, and mutually initiated sex. Emotional openness and the desire to express love go hand in hand with physical pleasure in bed.

SEX: GLUE OR SOLVENT?

One-way arrows of causality are generally passé in the new relationship science, so it is, of course, too simple to say that attachment shapes sex. In fact, they are a circle, one reinforcing or weakening the other. A strong emotional bond leads to good sex, which in turn leads to a still stronger bond, and so on. The reverse is true as well. A weak connection often leads to unfulfilling sex, which further weakens the connection.

This may be why unhappy partners in established relationships are so quick to cite sex as the major cause of their misery. Distressed mates assign up to 70 percent of their misery to sexual problems, according to sex educators Barry and Emily McCarthy of American University in Washington, DC. By contrast, less than a quarter of contented partners credit a good sex life for their happiness. Sexual dissatisfaction is actually a bellwether, the most evident sign of what's truly going wrong in the relationship: the unraveling of the emotional bond.

For more secure people, good sex can help overcome minor misattunements and even more serious difficulties. The emotional platform, the trust and safety built over months and years, is solid; synchrony sex helps glue any edges that are crumbling. The insecure are not so lucky. For the avoidant, there really isn't an emotional foundation to build on, and sealed-off sex never permits one to be constructed. In anxiously attached people, the foundation is flimsy. Good solace sex can help cover over cracks and gaps and keep the relationship steady for a while, but bad solace sex only widens the nicks and chinks, until the entire edifice tumbles down.

WHEN ATTACHMENT STYLES MEET

Unless one is masturbating, sex involves another person who has his or her own attachment style. The most relationship-damaging interactions occur between people who habitually avoid emotional engagement. Sex becomes an impersonal transaction, a bargain. When I ask such couples, "Do you make love?" they give answers like, "Yes. We schedule it. Every two weeks on Sunday at 7 p.m." Sex is merely scratching an itch in such instances; it does not promote emotional intimacy, since both partners are focused solely on satisfying their own sexual urges.

Other combinations are only marginally more successful—for example, an avoidant man with an anxious woman. In studies, avoidant men report having less sex as their partners express more anxiety and need. She pushes for more sex as reassurance, but her demands make him even more wary than usual, and he draws even further away. There is less sex and less satisfaction for both partners.

When two anxious people get together, there may be lots of sex as they both try to allay their fears. But they are so preoccupied with their individual concerns that neither is able to respond the way the other wants, and they wind up less trustful and more doubtful of each other's love. She: "If you loved me, you'd say *something* sweet while making love to me." He: "I'm so nervous about keeping my erection and performing, and also that you're going to reject me, that I can't say *anything*."

Sex best enhances relationships when partners' needs are complementary. Thus two secure people can pair rewardingly. So can a secure person with an anxious one. For example, Peter grew up in a large, close family on a farm, where it was natural and common to see animals mating and breeding. Mary, the only child of elderly parents, is a quiet, shy young woman with limited sexual experience. The two met at a local community college and quickly fell in love. Peter, with his confidence and warmth, gave Mary a safe haven where she could bloom sexually. And as Mary became more confident and secure, she gave Peter the affection and playful sex that he was longing for.

Recognizing that attachment shapes our sexual behavior changes our perspective on many issues. Once we understand love, these issues take on a new clarity.

FORSAKING ALL OTHERS?

Today there is hot debate in the media, on talk shows, and in academic journals about whether it's possible to stay with one person for a lifetime. The popular consensus seems to be that while it is a desired goal—according to a recent survey, 90 percent of U.S. teenagers hope to marry and stay with the same spouse "till death do us part"—it is impossible to achieve. People point to gloomy statistics as proof. Various surveys have found that nearly half of all U.S. marriages end in divorce and almost half of all American men and women cheat on their mate.

Observers cite numerous reasons why long-term monogamy is unrealistic. Familiarity breeds boredom, say many relationship experts. "Passionate love provides a high, like drugs, and you can't stay high forever," explains psychologist Elaine Hatfield of the University of Hawaii. Polygamy dominates in many cultures around the world, according to anthropologists. Constancy goes against the natural order; only 7 percent of mammals are monogamous, naturalists note. Among humans, men in particular are programmed to spread their genes around and assure survival of the species, or so evolutionary biologists theorize. In the light of all this, shouldn't we all just grow up and accept that promiscuity is natural and that romantic love has an expiration date?

Emphatically not! All of us may not be destined for a single, lifelong relationship, but we are naturally monogamous. Yes, that's right. Naturally monogamous. I hear gasps from an audience whenever I say this, but the evidence is solid: we are wired to prefer mating and bonding with one partner for the long term. Polyamory and short-term mating are not the strategies of choice for most humans, male or female.

What of the pessimists' points? If we look at them closely, they don't carry much weight. Divorce is actually dipping among those

younger than age fifty (and has always been lower in countries outside the United States, such as Canada). The figures on cheating
are often based on flimsy research and wildly exaggerated; reliable
studies indicate that only around 25 percent of men and 11 percent of women actually stray. Polygamy exists in more primitive
cultures mainly because men are few and because lack of education,
equality, and opportunity prevents women from supporting themselves and their children on their own. As for nature, 90 percent of
birds are monogamous. And although only a few mammal varieties
fall into this category, monogamy tends to be the rule in those who
must invest time and effort to ensure survival of their offspring
and the species as a whole. Among them are the California mouse,
pygmy marmoset, beaver, gray wolf—and humans. These mammals are all biologically wired to attach to those who depend on
them and to those upon whom they depend.

All of them also produce oxytocin, the neurotransmitter and
hormone that promotes bonding, both parent to child and partner
to partner (see Chapter 4). In humans, oxytocin surges through our
brains at moments of heightened emotional connection, such as at
breastfeeding and orgasm. Recent evidence shows that our lover
doesn't even have to be physically present to trigger a flood of oxytocin in our brains. We only have to *think* of him or her to be
inundated with it. This hormone also reduces the release of stress
chemicals and leaves us calm and blissful, further reinforcing the
bond between lovers. As I've mentioned, oxytocin is widely known
as the cuddle hormone, but scientists have another name for it: the
molecule of monogamy.

The clearest proof of this chemical's power in encouraging fi-
delity comes from studies of two species of voles, prairie and
montane. These little rodents differ in one major way: prairie voles
have oxytocin receptors in their brain; montane voles do not. Mr.
and Mrs. Montane mate, have young, abandon the pups after a

few days, and go their separate ways. Mr. and Mrs. Prairie, on the other hand, mate, have young, rear them, and stay together for life. When researchers boost oxytocin in the faithful rodents, they snuggle like crazy and almost groom each other to death. But when oxytocin is blocked, these loyal animals mate but do not bond, just like their montane cousins.

This was starkly demonstrated in one series of experiments. Researchers placed a mated pair of prairie voles in a cage and tethered the female. Then they opened a door to another cage, where an unfamiliar female scampered about. Did the hubby stray to go play with the intoxicating lady next door? He did not; he stayed with his wife. Then researchers injected him with a chemical that shuts down the brain's oxytocin receptors—and the prairie male became as shameless as his montane relative, copulating indiscriminately with both mistress and mate.

A recent study demonstrated for the first time a monogamous oxytocin effect in men. Neurobiologist René Hurlemann and his colleagues at the University of Bonn, Germany, administered an oxytocin nasal spray to a group of healthy heterosexual men, and forty-five minutes later they introduced the men to an attractive unfamiliar woman who moved around the room. Each man was told to indicate when he felt slightly uncomfortable at her closeness and when she was at an ideal distance. Oxytocin is known to promote trust, and the researchers expected all the men to permit the woman to come equally close. But that was not the case. The men in committed relationships kept between 10 and 15 centimeters farther away from the woman than did the single men!

Because of the link between oxytocin and sexuality, there is a natural propensity for sex to lead to bonding, thus inclining us to long-term relationships. Oxytocin has another fidelity-supporting effect, one that negates the "sex inevitably gets dull with the same partner" argument. In fact, in studies of cocaine addiction, oxytocin has been shown to interact with dopamine receptors in the reward centers of the brain and actively block habituation so that pleasure does not diminish. This seems to be evolution's way of ensuring that mothers and infants and adult lovers will find their interactions, including sex, infinitely and continually rewarding.

Of course, very little in nature is absolute and completely consistent. Oxytocin does not *guarantee* sexual exclusivity, so occasionally even the monogamous Mr. Prairie will mate with another lady. But then he rushes home to groom, sleep with, and protect his mate. Should we take this to mean that an occasional fling has a biological rationale—that it supports the old "it didn't mean anything" argument? No! The fact that we can occasionally get turned on by someone other than our partner does not mean that we are not suited for monogamy. We are much more complex than rodents.

Additional support for the idea that, in humans, mating and bonding are tied together comes from the recent work of Omri Gillath, professor of psychology at the University of Kansas, and his team. In one experiment, he had 181 heterosexual men and women between the ages of 18 and 40 each sit before a computer screen and look at twenty pairs of words that describe kinds of furniture—for example, "table-television" and "cabinet-chair." Their ostensible task: to press a number between 1 and 7 indicating how dissimilar or similar the paired furniture is. But before each pair of words appeared on the screen, there was a flash lasting just 30 or 50 milliseconds, much quicker than the blink of an eye, that the subject wasn't aware of seeing. Half of these subliminal flashes contained a neutral abstract image; the other half carried an erotic image—for example, male subjects got an attractive naked woman, female subjects a naked man.

Gillath then asked the subjects to fill in a questionnaire about how they would typically respond to their romantic partner in certain situations. The people who were subliminally "primed" for sex were more likely to check off intimacy-related statements ("I feel very close to my romantic partner") and positive ways of resolving conflict ("I try to cooperate with my partner to find a solution that is acceptable to both of us"). Men also checked responses that showed a willingness to make sacrifices—for example, to forgo seeing family or friends or indulge in hobbies or entertainments in order to maintain a relationship. Gillath's conclusion? Even lust, the slightest simple sexual arousal, automatically triggers attachment or bonding responses.

This fact, along with oxytocin, explains why adulterous one-night stands, swinging, and polyamory ultimately don't work so well. Wandering spouses may tell their mate, "It didn't mean anything; it was just sex," and polyamorous couples may set up boundaries and rules for encounters with others ("No kissing or cuddling; no meeting outside of set times"), but such assurances and restrictions are like moving chairs around on the deck of the *Titanic*. Chances are there will be a moment in the sexual experience when participants begin to connect emotionally—because we are set up that way. Nature has designed us so that physical closeness easily and inexorably slips into bonding and caring. Sex hooks us into relationships.

The truth is that we stray and have affairs not because we are all naturally inclined to have multiple mates but because our bond with our partner is either inherently weak or has deteriorated so far that we are unbearably lonely. We haven't understood love or known how to repair it. So, confused and lost in a world that sells sex aggressively as the be-all and end-all of a relationship, the only obvious "solution" has been to seek out new lovers to try to create the longed-for connection.

What about the argument that passion is impossible to sustain over the years? This is true—if we do not know how to invest in the security of our bond or if we only know how to have sealed-off, avoidant sex. More and more novelty is necessary to sustain attention if sensation and performance alone are the focus of intercourse. Then familiarity becomes the death knell of exciting sex. For secure partners, however, rigorous studies and surveys show that the thrill can last indefinitely. This excitement is not the explosive lust of first infatuation but a deeper exhilaration that rises from knowing someone profoundly. When I ask my client Jerry, who has been happily married for thirty years, about his sex life, he responds, "Do you mean the 'Oh, my God, this is fun, and this shows she likes me, and we are *so* hot' sex, like in the beginning of the relationship? Or do you mean the kind we have now, where we are really tuned in to each other—what I call 'soul sex'? It's still a total thrill, but it's a whole different kind of heat. This is like the morning sun."

Secure lovers have the capacity to be playful and adventurous throughout the relationship. This is borne out by a recent survey of sex in America by University of Chicago researchers who found that sexual satisfaction and excitement for both men and women increases with emotional commitment and sexual exclusivity. All this reminds me of my friend Mary, who dances tango, preferably with her husband of thirty years. She tells me, "I like to sometimes dance with other partners. But Marty and I, we have three decades of practice. We know how to help each other stay balanced, how to tune in, how to play. Dancing with him is delicious. He knows how I dance, and he is with me in a way that other social dancing friends cannot be."

If you are tuned in, every dance is different, even if your partner is the same. So, too, is every sexual encounter different, even if your lover is the same.

WOMEN AND LIBIDO

New science and the attachment perspective are also sparking a major revision of our views of female and male sexuality. A huge sexual problem today is lack of libido. Around 30 percent of women say that they have little or no desire for sex, even with a committed, loving partner. By contrast, only 15 percent of men report feeling little desire. Research indicates that we have not understood the nature of women's desire and how female sexuality profoundly differs from men's.

In men, a lack of libido is almost always linked to illness, such as heart disease or diabetes. But in most women, no physical explanation is apparent. Indeed, studies show that women often show the physical signs of arousal—their genital and vaginal tissues swell with blood and natural lubricant—but this excitement never surfaces to consciousness and felt desire. The paucity of explanations in this area has left many women feeling ashamed and guilty ("There's something wrong with me in my head") and men feeling mystified and frustrated ("I don't know how to help"; "What am I supposed to do for sex?").

Laura, 28, and Andy, 30, are newlyweds living in New York. Though Andy is always eager to hop into bed and make love, Laura can barely work up enthusiasm for even a once-a-month encounter. Both have been miserable, and now they've begun arguing. "I went to my doctor, and he told me that I had a sexual problem," she tells me. "He said that it isn't normal for a newlywed woman. That I should regularly feel this hot spontaneous lust for Andy if I loved him. The more ashamed and anxious I get about all this the less I want to talk to Andy about it. He knows that I had some bad sexual experiences as a teenager, but he says I should be over that by now and that there is something wrong with me. Now the more I avoid sex, the more insistent he gets about it and the shorter our lovemaking becomes."

What's going on here? For years, we've used a simple model to explain sexual function and dysfunction: sexuality is genitally focused and moves linearly from desire to arousal to orgasm to satisfaction. This model does hold true for men, for whom arousal is largely a physical experience triggered by visual cues. A man sees a woman in high heels and a tight skirt, blood floods into his penis, he feels the erection, and thinks to himself, "I am aroused. I want sex." But this model now appears to be all wrong when it comes to women. For them, sex is a more complex physical—and emotional—experience. And one of the heretofore unrecognized requisites for feeling desire, new research suggests, is feeling safe.

An exciting experiment by Omri Gillath and his colleague Melanie Canterberry appears to literally show this happening. They took brain scans of 20 female and 19 male college students who were told to look at abstract pictures and rate how much they liked them. The students were also told that they might be exposed, subliminally or supraliminally (that is, consciously) to other images, some of which might be sexy. When exposed to the sexy "primes"—naked pictures of the opposite sex, both men's and women's brains lit up. There were some very slight differences between men and women at the subliminal level in the parts of the brain that were most activated, but by far the most fascinating finding was that only in women, at both the conscious and unconscious levels, did the prefrontal cortex and other regions involved in making judgments and decisions illuminate.

"Women's brains clearly respond differently to sexual cues—the control regions of their brain always turned on in response to a sexual cue," says Gillath. "It seems like they have a natural tendency to pair safety concerns with lust. They are preoccupied with security, which makes sense—sex is simply riskier for them." Sex puts women in a very vulnerable position; they are smaller and weaker than men, often naked and on their backs. They have to overcome the natural fear that that helpless position induces. They appear to unconsciously ask themselves: "How sure do I feel about this person? Can I trust him?"

Women may also have an innate fear of becoming pregnant and so tend to be more vigilant. "They have more of a biological investment in potential offspring than do men," notes Gillath. He suggests that one of the reasons women may have more oxytocin receptors in their brains than men do is because they require more of the stress-reducing chemical: "Maybe they need lots of this to turn off fear and be able to become aroused and have sex."

Much more so than men's, women's sexuality appears to depend on the quality of the relationship rather than the intensity of the sensations in their skin. "It naturally connects to attachment and a safe-haven relationship," says Gillath. "And it helps us to understand why women, even after the Pill and feminism, still tend to be the gatekeepers in sex and men the initiators."

Psychiatrist Rosemary Basson of the University of British Columbia has posited a new model of female sexuality to replace the old linear genital model. It's a feedback loop and includes such factors as relationship satisfaction, emotional intimacy, and previous sexual activity, all of which influence sexual response. "Women often begin sexual experiences feeling sexually neutral," she observes, "and move into desire and arousal as a result of sexual cues from their partner. Their sexuality is often responsive rather then agentic. It is a reaction to a partner's sexual interest."

Recognizing that cues concerning safe attachment are fundamental to women's arousal and sexuality opens the way to new remedies. Previous medical therapies have failed here. Viagra, for example, does not work very well in women, likely because it increases genital blood flow rather than actually creating the experience of desire. Women with low libido particularly seem to need more sensual, teasing foreplay to cement their sense of security and move into the awareness of desire and arousal.

For men, that means overhauling their view of female sexuality and adjusting their verbal and physical approaches to make it apparent that there is desire for the person, not just for orgasm. This offers women reassurance. Cary tells me in a therapy session: "I can't get over it. All those sexual technique manuals I read. What a waste of time. She likes me to talk to her. I hate to be corny, but sharing my feelings seems to turn her on. Amazing. I used to keep asking her, 'Do you want to mess around?' I didn't get what a turn-off that was for her. I wanted her to show all this passion right off the bat. Now I understand that she wants to be held and whispered to and then for me to come on to her slowly. It works!" His wife, Jill, mutters, "Of course," and smiles. Jill offers that she does not always need an orgasm to feel satisfied with a sexual encounter. Although this puzzles Cary, it is not that uncommon.

The idea that women, newlywed or not, take longer to become aroused than do men and that they need to feel safe and be soothed first was completely new to Andy, Laura's husband. He lives in our sexually obsessed but relationally unaware culture, where avoidant sex is held up as the norm and even as the ideal. In our sessions, I was able to help Andy stop criticizing and pressuring his wife and share his own anxiety when she did not respond in bed the way he wanted. And as he did so, Laura was encouraged to come out of her shell and state clearly what she wanted and didn't want, before and during sex. "Please don't hold me down or put your tongue in my mouth," she told Andy. "That is instant alarm for me, and I just want to get away from you. I need gentleness first." Once she was able to ask for what she needed, Laura and Andy's sex life hummed along just fine.

Although men's sexuality is more direct, they, too, have a need for emotional intimacy. In therapy, they tell me that sex without it is "empty" and that with it, sex is better. For one thing, they can share their performance concerns and receive reassurance. The fact that nearly 60 percent of men stop using Viagra after the first prescription speaks to the fact that sex is a complex relational and emotional experience.

All this fits with what we find in EFT. When couples become more secure and satisfied with their relationship, the sex, even when we don't directly address it, automatically improves as well. And when there are specific sexual problems, we still begin by addressing the quality of the couple's connection. A secure base creates the sense of safety that all partners, and particularly women, need to move into engaged, flexible sexuality.

LOVE IN THE TIME OF PORN

There are not many models in our culture for learning the fine-tuned emotional and physical coordination that good sex—or, I should say, sex that is good for both partners—requires. Andy mostly learned about sex from his buddies and from magazines, books, TV, and movies. These cultural touchstones are still prime sources of information about sex for men and women. Today, for men especially, there is another source: the Internet.

What these sources, along with sex education in schools, all have in common is a concentration on the mechanics of sex. We don't hear much about the emotions that revolve around sex, the context in which it exists. One exception to this is women's romance novels. This genre, even though it represents by far the biggest segment of the book industry, hasn't gotten much respect (unlike thrillers, the genre that men love). When they first appeared, they were derided as bodice rippers for their depictions of corset-bound, submissive women enthralled by princes, pirates, and medieval warriors. But in recent years, these books have changed.

Today, they feature contemporary, capable, accomplished women, and they even tackle heavy issues, such as domestic violence. The important point here is that even in the most explicit of these books, sex nearly always occurs within a relationship, and the feelings of the partners are described.

One example of a new and titillating variation on the romance novel is the bestseller *Fifty Shades of Grey,* by E. L. James. This book and its two sequels are, on one level, saccharine clichés of the romance novel. A young, virginal girl, Anastasia, falls in love with a "prince," in this case billionaire tycoon and BDSM aficionado, Christian Grey. She agrees to enter his Red Room of Pain and become his "submissive," or sex slave. However, even though she is overwhelmed by lust—she seems to have multiple orgasms at the mere sight of his long, elegant fingers—she manages to keep her wits about her and alters the contract he asks her to sign, thereby setting limits on what she must accept. Of course, she discovers that his sexual proclivities are the result of his inner pain—he was abused as an adolescent—and she turns her wounded lover into someone who "has a wealth of love to give." This Cinderella story (or is it a Beauty and the Beast story?) is peppered with descriptions of spanking and other bondage, discipline, and sado-masochistic pleasures.

What does the popularity of this "romance" tell us? That women long for men who can masterfully take them over and so allow them to explore and surrender to their own and their partner's sexuality without anxiety or guilt? Perhaps. We know that active surrender is prevalent in women's sexual fantasies. Females do need to be somewhat still and cooperative for successful coitus to occur. So submission in women and assertiveness in men are important cues for mating. This book taps into this basic fact of sexuality and ties it in with the inherent longing for connection that is displayed in romance novels.

Fifty Shades of Grey waters down the woman-as-object aspect of pornography, which shows up in the formal contract that Anastasia has to sign. In it, her body is defined by her lover's desires and needs—she agrees to remain hairless and to have pain inflicted. The Internet's popular "three grandmothers" weren't titillated; their review included the phrases "Ouch" and "Never do anything that hurts." For me, the disturbing part of this book was the labeling of regular, non-BDSM sex as "vanilla"—the implication being that unless you include whips and go to the edge of pain, sex is less than satisfying; it's simply not "sensational" enough. In a society that does not understand emotional connection, we have to go to more and more extremes to drag our bodies into deeply felt physical excitement.

This book, with its insinuation that regular sex is by definition vanilla, is being made into a film. It will be another in a long line of movies that distort the link between emotional bonding and sexuality. Movie love makes no love sense. I had to scour my memory and go a long way back to find a handful of films that show an honest, accurate portrayal of good erotic sex within a developing or ongoing love relationship. Three old movies—and one more recent movie—stand out for me here.

Don't Look Now (1973) shows how attachment and sex work together. The romantic thriller centers on Laura and John Baxter, long-married spouses who are grieving the recent drowning death of their young daughter. They are in Venice for his work, and in their hotel room they go through familiar routines. They casually chat while each bathes; he lies naked on the bed reading papers. Slowly, she begins to stroke his flanks, and then they turn to tender, erotic lovemaking. They are long-term lovers, comfortable and easy with each other, and yet the sex has intensity. They know each other's body and how to give pleasure. The long sex scene is intercut with shots of the two after sex, as they get dressed to go out to

137

dinner. Sex is part of the continuum of their relationship, and the sex has the residual effect of bringing them closer. As they walk down the hotel hallway on their way to dinner, she is tight by his side, her hand wrapped around his arm.

The Big Easy (1986) portrays sex in a developing courtship. Remy McSwain is a cocky New Orleans detective attracted to Anne Osborne, a self-conscious district attorney investigating police corruption. With humor and charm, he helps her feel safe enough to take an emotional risk. In the sex scene, she is uptight and hurts him when she goes to touch his genitals. "Sorry, I never have been good at this," she confesses. As she continues, saying, "I can't do this; too nervous. I am embarrassed," he becomes tender and caring, and finally she is able to relax. As he slides his hand up her skirt, we see that she is now aroused. They are interrupted by a call, so they don't get to make love, but the scene ends with Anne, seeing that he has to go, saying, "It's okay; I never had much luck with sex anyway." Remy tells her, "Your luck is about to change," and she allows herself to whisper, "Come back," as he runs out the door. In another scene, she throws up after viewing a dead body and as he cares for her and helps her clean up, she kisses him and gets toothpaste all over his face. Later in the movie, he weeps in shame at his and his buddies' history of corruption, and she comforts him. After facing conflicts and danger together, the movie ends with their waltzing through their home, having just married.

A History of Violence (2005) has two contrasting sex scenes, one in which sexuality and secure connection come together in synchrony and one in which sex is detached, even hostile. The first scene shows a long-married couple, Edie and Tom Stall, in playful intimacy. She has dressed in her high school cheerleader outfit as a surprise and comes on to him. They roll about on the bed, at first teasing, and then serious. The scene shows loving connection unfolding into intense sexuality in an established marriage. The second scene occurs when the bond has broken and she no longer feels safe with him. He has shown himself to be capable of breath-taking violence, and she realizes that he is not the man she thought she knew during all their years together. He is now a stranger, and so, deprived of a sense of safety, she turns away from him. He reacts to this abandonment by overwhelming her and taking her on the stairs. Although she becomes aroused about halfway through the encounter, the end is brutal and sad. After sex, she pushes him away and goes to her bedroom, where she sits hugging her knees. Both of them are now alone. Their sexuality will not be viable unless they can find a way to reconnect and renew their bond. The final shot of the movie shows them looking at each other with naked vulnerability, and there is a sense that they see each other and will connect again.

Friends with Benefits (2011) demonstrates how hard it is to keep emotions and attachment out of the bedroom. After being rejected by their respective lovers, Dylan and Jamie are risk-averse: "I am going to shut myself down emotionally," he declares. "I'm done with the relationship thing," she announces. So the two friends decide to have sex with no emotional strings—"Two people should be able to have sex just like they play tennis." And indeed, when they hit the sheets, it sounds like a tennis match, with each yelling instructions and comments at the other. "What are you doing, trying to dig your way to China?...A little more to the left," she instructs him when he is giving her oral sex. He announces that he has to go pee, and this is hard to do with an erection, so she has to wait. Both are focused on the task of getting to the Big O. (This is funny and kind of endearing to watch; it reminded me of Woody Allen's quip "Sex without love is a meaningless experience, but as meaningless experiences go it's pretty damn good.") But soon they can't keep their emotions at bay. They care for each other in spite of themselves. She shows him her safe place, a rooftop hideaway.

He takes her home to his family, exposing his fear of heights, tendency to stutter, and love for his dementia-addled father. In his family home, they come together, and this time they have the kind of sex we call making love: they are caught up in sensuality and each other in such a way that there is no need for yelled instructions. The closeness scares him, however, and the next morning he turns away. Significantly, the movie ends with both characters taking an emotional risk by confessing their vulnerabilities to each other; she fears she is "damaged," and he fears needing her. Only then can they admit they are in love.

These four movies seem to be among a small number of exceptions to the flood of images of impersonal sex that pervades Western culture. But we have left out the medium that has been hailed as both the ultimate source of social connection and the ultimate threat to that connection: the Internet.

CYBERSEX

Of all the cultural changes that have occurred over the past few years, the most pernicious to relationships has been the gargantuan increase in the amount and availability of pornography. Porn has always been with us, but it was a trickle on the margins of society. Thanks to the Internet, it has become a mainstream flood. In the United States alone, 40 million people visit Internet porn sites at least once a month; 35 percent of all downloads are pornographic. Around 85 percent of users are men, and male youths under the age of eighteen are among the biggest consumers.

Porn is strictly genital sex. It completely divorces sex from emotional attachment, the springboard for optimal sex, which requires mutual engagement, attunement, and responsiveness. Porn reduces sex to sensation—intercourse and orgasm—and eliminates any connection to or respect for the user's partner. Imitating porn is a surefire recipe for being a lousy lover. It teaches men that all that is necessary for satisfying sex is a hard penis and a soft orifice. And by ignoring all the hard-won knowledge about female sexuality, it teaches women that they are mainly receptacles of desire and the servants of male arousal. The word *pornography* comes from the Greek *pornographos,* which translates as "writing about prostitutes"—the original exemplars of depersonalized sex.

The result of the porn glut is troubling: we are creating masses of avoidant men and anxious women. Women, intent on holding on to their men, are beginning to have their genital lips surgically enhanced to match those of women in porn. "The whole beauty of women's genitals is that they are all completely unique," a despairing sex therapist tells me. I've seen the consequences in my office, too. Over the past five years, more and more distressed couples have been coming in with pornography as a central issue in their relationship. Women complain of being deceived, betrayed, and humiliated; men protest that their actions are harmless and criticize their partners for being too uptight and less "sexy" than the women online.

"You've spent all our savings on this, and you spend all your time in the evenings surfing these sites," Marilyn says accusingly to Tony after she's discovered his penchant for Internet porn. "Now I get why when we've made love lately it's like you're somewhere else. Not with me. Where am I in all this? It's like you're having an affair."

"Such a fuss," Tony replies tersely. "How can I be having an affair when I have never left the house? It's not like I was really having sex with anyone else. I was only typing. It's just a fantasy." Perhaps, but it's a fantasy life that takes him farther and farther away from a secure connection with his wife.

Even worse, men are abandoning their real mates altogether in favor of the fantasy figures on the flickering screen. Men used to use porn to become aroused and then head for their partners. Now the porn itself has become the object of desire, as Wendy and Larry Maltz observe in their book, *The Porn Trap*. Porn offers immediate sexual arousal and release with completely compliant and ready women. "I've lost all interest in dating," says one fan quoted by Wendy Maltz in the magazine *Psychotherapy Networker*. "Porn is easier and more convenient than dealing with actual people."

Porn's supporters dismiss such criticisms and charge that opponents are simply pathologizing a legitimate form of sexual release. But many health professionals are now convinced that Internet porn presents an even more alarming danger: addiction. The word *addiction* comes from the Latin *addictionem,* meaning literally "a devoting." An estimated 6 to 8 percent of men are now thought to be "devoted to"—that is, dependent on—online porn. "After all," Tony admits, "escape into a high is always, three hundred and sixty-five days a year, just a click away."

Can you really have an addiction to what is, after all, the normal, naturally occurring process of sexual release? The elements of addiction are all there: a physiological "high"; the compulsion to seek out a "fix"; increasing tolerance of the "drug," which requires increasingly bigger fixes; a sense of deprivation when it isn't available; increasing preoccupation with the release offered; the investment of a large amount of time in pursuit of the fix; disruption of one's private and work lives; and so on. One element of classic addiction—that a person continues to seek out a fix to avoid the pain of withdrawal—is missing. But a newer view of addiction holds that addicts are first and foremost caught in a web of expectation. They anticipate pleasure from getting a high and an escape from anxiety and depression, and this anticipation creates a persistent state of semiarousal. My clients who are riveted to cybersex report feeling intensely and constantly sexual at a level that was unknown to them prior to using digital porn.

Porn slakes desire—momentarily. A screen-generated orgasm triggers a rush of "feel-good" chemicals, including endorphins, dopamine, and serotonin. It does not, however, discharge oxytocin, the attachment hormone, which produces consummate contentment and calm. Habitual users of porn soon find that they need more and more porn to get release. Porn simply makes you want more porn.

Recent research is showing that, in all addictions, overstimulation results in an excess of the reward hormone, dopamine, being released in the brain. To maintain equilibrium in the nervous system, the brain shuts down the receptor sites that take up dopamine, and response to dopamine slows down. This is like pleasure fatigue. As physiological tolerance rises, more and more stimulation—be it of cocaine, alcohol, prescription painkillers, or sex—is necessary just to feel normal, let alone high. In one study, Paul Johnson and Paul Kenny, neuroscientists at the Scripps Research Institute in Jupiter, Florida, found that overeating by rats resulted in dopamine deficits in the their brains and further compulsive eating.

The power of the Internet is unparalleled: it is unlimited in variety of content; it is private, anonymous, eternally accessible, and seemingly offers no real-life consequences. Tony's forays into cybersex escalated as his marriage got rockier and his work suffered (his boss caught him one day watching porn on the office computer and warned him that his job was in jeopardy). But he hems and haws about whether he has a serious problem. "I know I can stop it if I want," he tells me. "It started when I was down; I'd broken my leg in three places. And I just opened this e-mail. Now it's just that . . . I guess, well, I must admit that I miss watching it if I don't do it. I miss that high and the calm that comes afterwards. I guess it's true that I use it more and more, but so what? I feel so in control. I can play out my fantasies at the click of a button. My wife says that I have turned away from other parts of my life, that I am in love with porn. Maybe I have overdone it a bit. But our sex relationship was never that good. This stuff adds a kind of color to my life. It's like a secret feel-good pill in my pocket, and I guess I feel more sexy as a person knowing that I can watch as soon as I get home. And now Marilyn is all angry. I never saw it as anything to do with her, really."

When Tony describes his attachment to Marilyn, it is clear that he has an extremely avoidant style. Their courtship was all about "having fun," he says, and what tied them together was their mutual love of outdoor sports. When I ask whom he feels close to, confides in, or turns to for support, he looks at me blankly. "Adults need to stand on their own two feet," he replies. "The way I was bought up, you were sent to your room if you got upset. You have to learn how to deal with things yourself." There is good evidence that avoidant folks are more susceptible to addictions in general. If you cannot find your way to healthy attachment, you go in search of a substitute.

Porn addiction is a perfect example of the consequences of cutting off sex from attachment and connection with others. Sex and attachment are meant to go together. Most addictions are, at base, desperate attempts to find a substitute for secure attachment to others. But such substitutions cannot satisfy, and they are destructive to health, happiness, and even, ultimately, to sexual functioning. As men become accustomed to porn's high-octane stimulation, they become desensitized to the pleasures and the physiological highs of regular sex. When they are with a real partner, they find themselves unable to become aroused. Urologist Carlo Foresta, professor at the University of Padua, has found from his surveys that 70 percent of the young men seeking help for sexual performance problems admit to routinely using porn. He suggests that a numbed-out sexual response system resulting from obsessive use of porn, not performance anxiety, is now one of the main causes of erectile dysfunction (ED). Clinicians are finding that if men can abstain from porn for a period of time, their physiology eventually recalibrates, their sexual performance improves, and their libido rekindles.

In the end, Internet porn devastates our capacity for close relationships and good sex. It promotes loneliness and isolation and infuses a person with shame and despair. Porn devotees are left with a broken and fragmented sexuality, in which emotion and the erotic are separate and never integrated.

It is only when Marilyn walks out on Tony that he begins to recognize the price of what he calls his hobby. He goes into individual therapy, and he and Marilyn come back to couple therapy. Letting go of porn and turning back to the complexities of sex with a real person is a huge challenge for Tony. He is still working on it.

Science today is offering us a new understanding of sexuality: mature sexuality grows from and flourishes in a secure sense of attachment to others. As the actor Peter Ustinov quipped: "Sex is a conversation carried out by other means." Where there is no conversation—no emotional connection—the consequences are dire. But when we bring attachment and sex together, there is nothing better, and it makes perfect love sense.

EXPERIMENT 1

Famed sex researchers Masters and Johnson said that there was a simple biological motive for sex: an "inborn drive to orgasm." In fact, though, there are many reasons why we have sex, and our attachment style shapes these.

Try looking at your own motives for having sex. Assuming that you are not trying to conceive a child at the moment, imagine the last couple of times you initiated an encounter or responded to a sexual overture.

On a scale of 1 to 10, where 1 is not true at all, 5 is moderately true, and 10 is completely true, rate the importance of these factors in your sex life:

I want to get close to and feel connected to my lover.

I want the turn-on, the thrill, and the pleasure of touch and sex.

I want the tension release, and sex helps me to let go of stress.

I want to feel special to my partner and cared for by him or her.

I want to show my love and have my partner feel special and cared for.

I want to feel good about myself and know that I am potent as a sexual person.

These motives are the most obvious. Maybe you have a special reason for seeking sex right now that is not on the simple list above.

Sam, who is recovering from his wife's infidelity, says, "I want us to make love because then I know she is mine. It's like when we were dating. She becomes my woman. It is like I am claiming her. I feel safe then."

Discuss with your partner your motives for having sex. The best sex combines all of the above factors for both partners. If you find that you and your partner are focused only on one goal, or that each of you has a different goal, explore that together.

EXPERIMENT 2

Secure attachment allows us to engage fully in a sexual experience. To examine your sense of security, sit quietly and bring up images of two or three moments when you felt really loved by others, and then imagine yourself in a sexual situation where you feel entirely safe, cherished, and accepted.

Ask yourself: "Feeling this way, what would I do or ask for that is different from my usual responses or routine in sex?" Another way to think about this is to ask: "If I could be emotionally as well as physically naked in bed with my lover, what would I do that is different?"

State this "discovery" in a clear short sentence, and share it with your lover.

PART THREE

Love in Action

CHAPTER 6

Love across Time

Love doesn't just sit there, like a stone, it has to be
made, like bread; remade all the time, made new.
— *Ursula K. Le Guin*

Embrace. Kiss. Fadeout. With few exceptions, an ecstatic
clinch marks the end of most romantic movies and reality
TV dating shows. (Have you watched *The Bachelor* or *The
Bachelorette?*) From then on, the implication is, the couple lives
happily ever after. But that is fantasy, of course. (TV's "true love"
pairs usually break up within months.)

We may like to dream that relationships are fixed at their most
joyous state, but we know better. Relationships are not static,
frozen-in-time unions; they are living, breathing organisms, reacting through the days and years to the outside world and their own
internal dynamics.

Relationships are tested constantly. We are well aware of the trials caused by accidental or intentional cruelties, such as illness or infidelity. But less well recognized is the profound challenge presented by even the most desired and welcome events. What the revolutionary new science is teaching us is that long-term relationships go through distinct periods—an initiation phase and three major subsequent stages—and that within each are critical transitions that shake every couple, even the most secure and serene.

The prelude to every relationship is what I call the Spellbound phase, during which two people become infatuated and increasingly obsessed with each other. When the two shift into a more explicit dependency and commitment, they enter the first stage of a relationship, which I call Formal Bonding. This typically occurs between one and two years into dating. The second stage, Parenthood, centers around the appearance of a couple's first child. This is an especially trying time for women, many of whom become deeply unhappy and even clinically depressed. A couple enters the third stage, Mature Love, usually when the last child is ready to leave home. Another stressor, retirement by one partner, may occur at the same time or later.

These are crucial transitions; partners' lives change dramatically and unpredictably as new challenges arise and different needs come to the fore. The smooth, known path suddenly becomes uneven and strange. While lovers may experience intense joy, pride, and excitement, they also grapple with massive stress and uncertainty. A couple often falters at such times. The usual explanation is that the general strain of these life transitions has proven too much. But more is at work here than has previously been understood.

These relationship shifts are actually potential bonding crises in which our need for connection and the nature of our bond is the core issue. A couple's emotional balance wavers; partners' faith and trust in each other often come into question. At such times, the bond has to be reshaped and renewed, or it may break under the weight of the new reality and each partner's changed expectations. The more we understand these stages and shift points and the relationship needs that arise from them, the more equipped we are to deal with them.

SPELLBOUND

At the very beginning of love there is infatuation and obsession. We tend to think that this is strictly the result of sexual desire. But right from the beginning, there is also emotional yearning. Indeed, as psychologist Paul Eastwick of the University of Texas, Austin, observes, passion is best defined as a combination of sexual connection and attachment longing.

A budding relationship is fraught with tension and anxiety. We whisper to ourselves, "Does this person want me? Am I going to be rejected?" The longing and apprehension push us to take risks, to reach out and move closer. Our anxiety is soothed as we get positive responses from this person, and gradually he or she becomes what John Bowlby called "irreplaceable." The process of feeling anxious and vulnerable and finding that another can and will respond is the basic building block of love.

In movies, protagonists often dislike each other at first sight, but once they slay a few dragons together and discover solace and protection in each other, they realize that they are in love. Psychologist Lane Beckes of the University of Virginia, Charlottesville, has found that, indeed, any kind of threat automatically turns on the attachment system, calling up our need for comfort and making others who are potential sources of this comfort more attractive. Beckes assessed 48 students on their level of attachment security and then asked them to view on a computer screen brief clips of four smiling faces of men and women that were paired with subliminally flashed pictures of either neutral objects, say, a rolling pin, or disturbing images, such as a striking snake. Then students were instructed to press a key if the letters that flashed on the screen made up a word.

Researchers found that the students were much more likely to recognize attachment-associated words, such as *nurture, comfort,* and *trust* just after they saw the snake image. In addition, those who were assessed as insecure were better at identifying such words as *rejection* and *vulnerable*. Students also rated the pictures of faces as more attractive, warm, and likable after the scary images.

Anxiety and threat automatically call up the need for comfort and prime us to find security in another. If someone is there at a vulnerable moment, we begin to bond, and every risk we face together thereafter strengthens the sense of connection.

FORMAL BONDING

Many romantic partners break apart when one person starts to ask, "Are you there for me?" and cannot get a clear answer. It is one thing to accept you're having a casual amorous adventure and another to face up to another person having a hold on your heart. Then you question how much you can really depend on that person, how strong is the devotion on his or her end. Many couples founder at moving into an explicit commitment, which frequently takes the form of a willingness to marry.

But is formalizing a bond really such a significant shift, such an emotional event? This may strike many as a silly question, given that so many couples today live together before marriage. About 41 percent of U.S. couples now cohabit before they wed, compared with only 16 percent in 1980. So how much of a change can there be after an official ceremony? A lot, researchers have found.

Living together may fully acquaint you with someone's everyday habits and likes and dislikes—he drops his dirty laundry on the floor or in the hamper; she wants the right or left side of the bed—but it often stops short of complete emotional linkage. It's like bouncing on the diving board but not plunging in. Moreover, cohabitation seems to have a hangover effect. Data show that couples that have lived together are more likely to be dissatisfied with marriage and to divorce. Why this is so is unclear, but it may be that couples who live together have more general reservations about marriage, more ambivalence about long-term commitment, and are less religious. Religiosity seems to encourage partners to wed and, when problems occur, to struggle to stay married.

Marriage allows full emotional commitment in two ways. It formally transfers attachment from one's parents to one's partner. It also allays anxiety about attachment and lays the groundwork for a long-term bond to grow. As a colleague said to me, "Standing up in front of all your close family and friends and putting a ring on each other's finger is a statement of your intention to be this person's love and home."

The significance of getting married has emerged, with a certain ironic clarity, in the fight gay couples have been waging to be able to legally marry. Many conservatives who oppose gay marriage still assume that the gay lifestyle is naturally promiscuous and that gay men, especially, are uninterested in long-term commitment. But recent surveys corroborate the fact that gay youth, who now can be "out of the closet" and thus have no need to resort to the casual mating that often accompanies a hidden life, overwhelmingly want to form stable, lasting bonds with their lovers. They don't want to just live together or have civil unions, they want to get married. "I don't want to be Stuart's 'friend' or his 'live-in lover,'" Owen, a marketing consultant in his thirties, tells me. "I want to be his spouse—to have that commitment and for us to take that leap, to say we will be together no matter what."

The early months of marriage are emotionally wobbly for nearly all couples. The wedding itself is often a roller-coaster ride of highs and lows. The bride is immersed in preparation for what marketers have cleverly labeled "the most important day of your life"—finding the gown and the venue for the ceremony, choosing the bridesmaids and their dresses, and ordering the food and cake. And that's before the emotional demands from parents, relatives, and friends pour down. There's little energy left for the groom, who often feels pushed aside and neglected, a postscript to the event. The TV series *Everybody Loves Raymond* captures these feelings perfectly in the episode when Debra, upon becoming engaged, pulls out an inches-thick album, a wedding planner she's been putting together since the age of twelve. "But you didn't meet me until you were twenty-two," says a shocked Ray. "Well, you're the last piece of the puzzle," says Debra.

Once the acute turmoil is over, couples undergo a more subtle but more formidable emotional shift. The rules for making marriage work were pretty clear-cut just fifty years ago. The husband was the breadwinner; the wife was the homemaker. I remember my granny telling me when I was very young that when I grew up I had to tell a man how clever he was, keep the house clean, and know how to make a good steak-and-kidney pie. "But I want to have adventures," I responded. "I don't want to keep the house clean, and I don't like steak-and-kidney pie." I never did learn to make the pie, but I did learn to tell a man he was clever, and I did get married, perhaps because the rules changed!

We tend to forget just how significantly expectations have altered. It wasn't that long ago that marriage was considered an alliance aimed at fortifying defenses, preserving wealth, and achieving financial security. Now marriage is seen primarily as an emotional venture, a commitment to the creation of a very particular kind of bond. In fact, in the United States, "emotional support" and "friendship" have replaced the rearing of a family as the central motive for marriage, according to a survey by the Pew Research Center, in Washington, DC.

Consequently, whatever disturbs the emotional relationship is paramount. A 2000 survey by the Center for Marriage and Family at Creighton University, in Omaha, discovered that the number-one problem reported by modern newlyweds is balancing job and family; the second is frequency of sex. Studies show, however, that even though satisfaction with the relationship tends to decline somewhat in the first years after marriage, the sense of security increases. Partners are less worried about abandonment and more comfortable depending on their spouse.

"It's not all roses and bells, like it was when we were dating," acknowledges Samuel, 29, who has been married for three years. "We fight more than we used to when we lived apart, and sometimes we have different ideas about what a husband or a wife is supposed to do. But for me, I know more than ever that she is the one I want to be with, and I think we can make it, even if we face problems. She is my wife. We count on each other. We are just going through a rough patch right now."

Newly married folks are more accommodating, according to a study by psychologist Scott Hall of Ball State University, in Indiana. They tend to minimize problems and anxieties and excuse harsh actions and words with statements like "He didn't mean to say that" or "Maybe I'm being too sensitive." These couples focus on the positive aspects of their new situation.

One consistent research finding is that the more insecurely attached people are, the shorter their significant relationships tend to be and the more likely they are to divorce. This fits with University of Texas professor Ted Huston's landmark study of couples married for five years. He found that the most important factor in predicting a marriage's collapse was not the amount of conflict present but rather the couple's lack of emotional responsiveness, a classic sign of insecure attachment.

PARENTHOOD

A greater emotional jolt occurs when two become three. Scores of studies conducted since the 1980s document the fact that relationship quality plummets when the first little one arrives. A 2007 study of 130 young families by psychologist John Gottman of the University of Washington, Seattle, found that in the three years after a baby was born, marital satisfaction dropped significantly in *two-thirds* of the couples.

Why is this? For both partners, there is less money, less sleep, more tasks, and more conflict over how to parent: new moms and dads have eight times as many arguments as childless partners. Parents suddenly see their roles shifting. Men may begin feeling hugely responsible for their expanded family's financial well-being and, as a result, pitch themselves into work. Women become the chief baby caretakers and can find their home workload tripled, according to the Center on Population, Gender, and Social Inequality at the University of Baltimore. "I have to get that promotion," say new fathers, and they put more hours into the office just when their wives are demanding, "I need you to come home at five o'clock and relieve me. I'm going crazy being home with the baby full-time." And there is no outside relief. Couples used to live close to relatives, now they often dwell states away. Grandpa and Grandma are only occasional visitors, not ever-ready babysitters.

New parents can soon wind up feeling isolated from each other. The most significant drop in relationship satisfaction seems to happen about one year after the first child is born. Partners find that they have little energy to put into intimacy and sex. They are out of sync. New mothers appear happy to have sex every few weeks, while new fathers want it three times a week. Some of this is probably the result of women's fluctuating hormones; oxytocin surges during breastfeeding, bonding mother and child, while testosterone and other desire-fueling hormones take a dive.

Terry, 35, a restaurant manager, who wanted desperately to be a dad, finds himself telling his wife, Chan, "Look, it's hard to say this because I feel like a wimp, but you are in love with the baby! And I am, too, but you are breastfeeding and snuggling for hours every day. The closeness we had in sex has kind of disappeared; for you it is almost like a chore. I can't help feeling left out and kind of deprived. And maybe this is ridiculous, but I watched my brother's marriage come apart when his kid was born, and I find myself getting freaked out and resentful. I don't want that for us. I am working harder at work, too, because suddenly I feel like the family provider. I don't want to pressure you. You're a great mom."

The strains of parenthood are so well known that only about 40 percent of Americans now believe that children are crucial to a successful marriage, reports the Pew Research Center. And couples with kids are more likely to divorce in the first seven years of marriage than are childless pairs.

The decline in marital satisfaction after baby's arrival is about twice that reported by new parents in the 1960s and 1970s, reflecting the new centrality of emotional connection. Loss of time and intimacy is more troubling because it is more valued. Partners are no longer expected to be just dependable teammates, they must be intimate loving soul mates as well.

"You think being a parent is changing a diaper once a month," Cindy says accusingly to Dan. "You leave everything to me, and then when I am exhausted, you want sex."

"You are right," replies Dan. "I probably don't do enough as a dad. But I am still struggling with how *we* have changed. Is there a *we* anymore?"

The Role of Attachment

New parents who have difficulty after baby's arrival tend to be the ones who were having trouble creating a secure bond before. Jack and Naomi had struggled from the beginning. Naomi is anxiously attached; she had an abusive first marriage and was reluctant to trust any subsequent lover. She insisted that Jack prove his love by moving across the country to be with her. After much arguing, he had agreed, but had then withdrawn into his work, trying to regain the success he had been forced to give up.

Their disconnect surfaced in the delivery room. Naomi was attended by her doctor, a midwife, and a doula. In the midst of a contraction, she said to Jack, "I want you to come and put your arms around me." Jack turned away and responded with a flat "no." Naomi felt wholly abandoned. "I felt replaced," Jack explained. "She had all these people taking care of her. I felt superfluous, incompetent, and unnecessary. I knew that whatever I did was going to be wrong, and I was just in the way. I guess I got scared—the birth wasn't going all that well, and I couldn't do anything."

Jack and Naomi's sense of isolation and rejection has not abated. Three years later, they still fight about what happened in the delivery room and about events since then, when Naomi has felt dismissed. In one incident, Naomi was deeply worried that her newborn son was throwing up too much and becoming ill. Jack responded by searching the Internet and presenting her with statistics showing that this was common among infants. He called her anxiety "ridiculous." She was not reassured. The distance between them widened, and conflict escalated.

If this couple had developed a more secure bond before moving into parenthood, they would have been better able to heal the original injury and to tolerate each other's different ways of handling baby-related anxiety as well as Jack's lapses into coldness. The effects of attachment insecurity always become more apparent when people face challenging situations.

Avoidantly attached mothers and fathers are apt to be ambivalent about becoming parents to begin with. Often they do not actively choose parenthood; they agree to it to placate their partner, or it results from contraceptive mishap or absence. They find tending to a child unsatisfying and frustrating when it interferes with their personal interests and activities. Their complaints are not centered on a partner's lack of support—they dismiss their need for support most of the time—but rather on their own irritation and discomfort. "I used to go the gym every day," protests Carl. "Now I can't get out of the house."

"We used to go out for brunch on Sundays or to a movie with friends on the spur of the moment," complains Sylvia. "Now on Sundays we're always taking walks in the park with the baby. And we can only see friends with a lot of planning. Which means almost never."

Avoidant people are less available and responsive to their partners and, as hundreds of studies of parents interacting with their infants demonstrate, to their children as well. The weak bond makes transition periods especially difficult for couples and also generates a less effective parenting style, which can adversely affect the growing child.

Couples with a secure connection are not immune to trouble during life transitions. They, too, can feel overwhelmed by the challenges, miss each other's signals for help and reassurance, and move into negative cycles, for example, habitual blaming followed by defensive withdrawal. But they are better able to tolerate and recover from the inevitable periods when a partner is less responsive and available. They have faith in their partner's love and their ability to renew connection when opportunities occur. They will even *make* the opportunities occur. John Gottman calls such flexible pairs "master couples" of the transition to parenthood.

Cindy, for example, hears Dan's message that he needs her attention. She tells him that if he can arrange to come home early a few days a week and care for the baby, she will take a nap instead of doing laundry. That way, after the baby goes to sleep, she'll be rested and alert and the two of them can enjoy dinner conversation and the evening together.

Postpartum Depression

To what extent are partner-bonding issues related to the rise of clinical depression in new mothers? Postpartum depression is estimated to occur in between 7 and 15 percent of women, and the less severe but still distressing "blues" in between 30 and 50 percent. Sarah, 35, tells me, "I have never felt so overwhelmed in my life. I am so scared of not being the perfect mother and I am so exhausted and the demands are so constant. And Gerry just doesn't seem to understand. I feel more distant from him just when I need to feel really close."

Depression in new mothers traditionally has been viewed as the result of the powerful hormonal changes that occur during childbearing and birth—that is, as a purely physical medical issue. It's only recently that researchers have begun considering the impact of the attachment relationship between partners. John Bowlby himself pointed out that uncertainty and stress heighten our need for a "safe-haven, secure-base relationship" and that perceived abandonment and rejection, or even a lack of comfort and support when we need it most, naturally generate depression and despair.

We now have strong evidence that, although hormonal changes play a role, depression is deepened and perpetuated by relationship anxiety and distress. This is particularly true for anxiously attached women, who tend to be highly sensitive to their partner's actions and so are more likely to perceive that their need for support is not being met. They report more anguish and anger and have a higher incidence of postpartum depression than either their avoidant or securely bonded peers. Studies suggest that many of these women may also be depressed before pregnancy and motherhood. In addition, anxious women see themselves as less competent, and this affects their ability to adapt to their new role as mother.

Highly avoidant women suffer less depression than do anxious women after baby's appearance. They tend not to enjoy caring for children, and so are more detached as mothers. They endorse harsher methods of discipline and expect their children to become separate and independent at earlier ages than do mothers with other attachment styles.

The women who fare best when the "blues" or depression strikes are those who are securely bonded with their mate. They are less hesitant to ask directly for support and caring and recover more quickly. Even when depressed, these mothers can parent effectively and sensitively. Carolyn and Philip Cowan, psychologists at the University of California, Berkeley, followed 96 couples from pregnancy to their child's enrollment in kindergarten and beyond and found that depression only seriously compromised a mother's ability to be warm and responsive to her children when she felt that her marriage was also in jeopardy. Secure bonding protects us—and our offspring—from negative emotions that are triggered in times of stress.

Depression after a baby's birth occurs in men as well as women, though it is less recognized. This may be because of the delay in its appearance. Women can sink almost immediately into melancholy; men typically become ill two to three years later. Unlike women, men experience no hormonal turmoil; it's strictly an emotional problem. Researchers now link it to distress in the couple's relationship. A toxic feedback loop develops: the frayed bond leads to conflict, which in turn leads to depression, which further weakens the bond, and so on.

Treatment has typically focused on the individual parent, usually the mother, but new research indicates that it would be more effective to treat the couple. Shaila Misri, clinical professor in the department of psychiatry and obstetrics at the University of British Columbia, studied 29 couples in which the new mother had been diagnosed with postpartum depression. Half the mothers met alone with a counselor for four sessions to discuss how to cope with a new baby. The other half attended sessions with their baby's fathers. There they were encouraged to talk about how to handle the baby and household chores and also how to talk to each other about these issues. At the end of the intervention, the couples in which the mothers were seen alone reported decreased relationship satisfaction. The "together" group, by contrast, showed an increase in happiness as well as a lessening of the mother's depressive symptoms, such as crying and hypersensitivity.

Another program, Bringing Baby Home, developed by psychologists John and Julie Gottman, has had similar results. During the two-day course, couples perform communication exercises and view videos that demonstrate how to care for and play with a baby. Couples enrolled in the program were happier with their relationship and had fewer signs of depression compared to couples who did not attend.

Strengthening the bond between spouses is the best way to ease the transition to parenthood. A secure relationship not only makes partners feel safe, it also instills confidence in their individual abilities. Such faith when we are entering a new stage in our relationship and our life can be amazingly effective in enhancing our ability to cope.

In our last session, Elaine told her husband, Mark, "I was used to feeling so sure of myself at work. But suddenly, when Joey was born, I had no confidence at all. I felt so overwhelmed when he wasn't sleeping and was so difficult to soothe. I didn't know how to do this—this thing that all women are just supposed to know by heart. I was failing the Madonna test. But then you gave me all this emotional support. You didn't tell me what to do. You told me that you saw how exhausting it all was and that you were sometimes unsure, too, about how to be a good parent. Best of all, you told me that you thought I was a great mom. You appreciated me for working so hard to figure out how best to comfort Joey even when he wouldn't suck properly, and you felt lucky to be married to me and to have me as the mother of your son. You told me that you knew I could do this and we would figure it out together. That changed everything for me. I would think about you saying that when I was alone in the house, exhausted, and Joey was crying."

There is, of course, even more at stake here than the quality of the relationship between two people. How a couple deals with becoming parents and functions as a caregiving team inevitably influence their children's emotional and mental health. Irritability and withdrawal, both part of depression, will threaten any child's sense of secure connection. Depression, particularly in a mother, is one of the strongest predictors of emotional and mental problems in children and adolescents. As such, how a couple navigates the shift to parenthood shapes future generations and our society as a whole. Indeed, in this as in many other concrete ways, the strength of our romantic relationships impacts us all.

MATURE LOVE

As we grow older, relationships have to accommodate to critical challenges: children leaving home, spouses retiring, and partners aging and becoming physically frail. A more secure bond improves our ability to deal with these transitions and to renew and grow our relationships.

The Empty Nest

Children ultimately leave home, of course. For some parents this transition is painless and even positive. They can celebrate this shift as the beginning of a second honeymoon, a chance to celebrate their bond. Marta and Ken were sad to see their last child choosing to take his first job in a faraway city, but they decided it was a chance to look at their relationship and maybe even plan to renew their vows in the church down the road, where they had married thirty years before.

But for others, moving into this stage of love is rife with grief, loss, depression, and marital conflict—so much so that many couples wind up divorcing. The empty nest syndrome, as it's been dubbed, has usually been ascribed to a parent's—commonly the mother's—overinvestment and overinvolvement in the offspring. The attachment lens, however, reveals a new explanation: children flying the nest exposes a big emotional gap between partners.

For many couples, kids have been the bridge over the attachment abyss in their marriage. For years, they've been joined in parenting but not much else. Once the buffer of the children is gone, their disconnection becomes overt and inescapable. They find that they are incapable of reaching out to each other to deal with the loss of the parenting role and the everyday intimacy with their children. They are not each other's safe haven.

Kali, 54, is contemplating separating from her husband after thirty years of marriage. "From the beginning, we had lots of problems," she observes. "I always felt kind of inferior to Frank, so I would hide and withdraw. But when we had the kids, it really pulled us together. We were a great team, and that was the basis for our closeness, I guess. We were twenty-four-hour-a-day parents.

I was a mother first and foremost. But then our youngest left for college, and suddenly I had no idea what to do with myself or what to say to Frank. There was this empty space in the middle of our relationship. I felt so lonely, and I realized how much I had counted on the kids for closeness." Kali turned to her husband for help with this loss of identity and connection, but he was unresponsive. "I needed to talk. I needed his help, but he just went off, back into his career. Then I got good and mad."

Women like Kali not only get emotional sustenance from their children to compensate for the lack of nourishment from their spouse, they also often actively tamp down their own anguish to keep the family intact for their children. After years of such bottling up, when the last child is launched, they erupt. Staying together for the kids is hard at the best of times, but it is impossible once the kids have gone.

Spouses who are securely connected are less threatened by and better able to cope with the loss of the parenting role. They are able to seek their partner's support and in return be responsive to their partner's needs. They can go through this transition as a couple. For them, this period can be an opportunity to not only affirm the bond between them but also use their connection as a secure base from which to vault into a new life and develop fresh interests of their own.

Claire, 53, an attractive but shy mother of three sons, is married to Simon, 55, a busy lawyer. Claire homeschooled her children, all of whom have learning disabilities. Her youngest, Todd, recently went off to join the navy. Simon, his father, was ecstatic. For him, this event signaled success as a parent and a new freedom. But he was concerned about how Todd's leaving was affecting his wife. In my office, they discussed what was happening.

> *Simon:* I am worried about you and about us. You mope
> around all day and tidy up the house when it doesn't

need tidying. I don't know what is happening with you, but you seem very distant, and we don't seem to share or cuddle. I want to help you cope with the kids leaving. It must be hard. I think you should start going to the gym. I want to help.

Claire: I don't want to go to the gym. I am fine. [She sounds angry, so I ask her about that.] I guess I am. I have lots of conflicting feelings here. I have looked forward to not being such a hands-on mom, and now that that time is here, I feel lost. And you [she points at Simon] keep trying to manage me and cheer me up, tell me how great everything is, how we can travel now. Everyone keeps telling me that I should feel really good that Todd got into the navy and all my boys are launched. Maybe I should. So I keep quiet.

I ask Simon how he feels about Claire's silence.

Simon: I don't like your quietness. You miss the boys, but I miss my wife. I miss you. I don't know what is going on with you. So I suggest stuff like the gym. And you just get irritated with me. It's been weeks now. Where are you, Claire? Where did you go?

Claire: [She smiles and tears up at the same time.] I just need to be sad for a bit, and my sense is that you can't hear that because you are relieved that it's just us at last. So I go be sad by myself. I liked being a mom, even when it was hard. [She cries.] Now they don't need me, and it feels bad. I know there are other things I can do—another page to turn—but right now...

Simon: [He leans forward, his face soft.] You are a great mom, sweetie. Your kids will always turn to you. In a way, I am relieved, but I miss them, too. I don't want you to feel so sad. It worries me.

Claire: I guess I just need some comfort from you and some time. I can't just turn around and remake my life. I feel a little raw here. And unsure of who I want to be now. I need a little help with my feeling sad. I can't just act like nothing has happened.

Simon: [He reaches for her hand.] I will help. Whatever you need. Sorry if the "go to the gym" idea was off. I was trying to help. I guess you need to grieve a little here. To be sad. I will be here. [After a long silence.] And I hope that you will get to the place where you are happy with us just being us, a couple, and be happy with that. Do you think?

Claire: [She laughs.] Oh, yes. You don't have to worry about that. We will be fine.

She looks at me. I sense her wondering about my silence. I tell her that they don't need me at all. They are doing just fine helping each other find their feet and move off into a new dance together.

A few months later, Claire went back to school part-time and took up photography so she could take pictures on the trips she and Simon were planning. She told me that she had fallen in love again with the guy she had married twenty-seven years earlier and that this was strange but good.

Retirement

The Beatles song "When I'm Sixty-Four" highlights yet another transition that we are only now beginning to understand: becoming elders. Today, our life expectancy has increased; many of us will live well into our eighties and even nineties, and some will make it to one hundred or more. And we will need to love and be loved during all those years.

Traditionally, researchers have believed that marital satisfaction follows a U curve: it's high at first, dips through the child-rearing years, and then rises again once the demands of parenting and work have lessened and a couple has more time for each other. There is some evidence that older couples do tend to fight less, have fewer negative emotional responses, and show more affection, even during arguments.

Yet the facts belie the U curve model. Unlike the divorce rate for younger age groups, which has been holding steady since the 1980s at around 45 percent in the United States, the rate among those age fifty and over is on the rise. Tipper and Al Gore, who split after forty years together, are the best-known example of "gray divorce."

Women in particular are not as willing as they once were to stay in empty-shell marriages, according to a 2004 AARP study of divorce at midlife and beyond. Sixty-six percent of the splits were initiated by women, and 26 percent of the men stated that they never saw the divorce coming. Among men and women ages 65 and older, the divorce rate has at least doubled, and we should expect to see more breakups as the population ages: 13 percent of Americans are now over 65; by 2030 the figure will be 19 percent. This is in spite of the fact that there is evidence that partners tend to mellow, fight less, and generally become nicer to each other in their later years.

Various explanations are offered for gray divorce: we stay healthier longer; baby boomers value personal happiness; women now have the financial means to live on their own. But these are what permit couples to separate—they aren't the root cause. The foundation of contented, sustained relationships is the faith that your partner is there for you. And that trust can be brutally rocked in the elder years.

170

The first shock is retirement. A partner's leaving the workforce can prompt many of the same feelings as the last child leaving home—and reveal the same emotional breach. Partners are thrown together after years of spending considerable time away from each other, and their goals and needs may be poles apart. They can feel like they're living with a stranger.

John, 66, has just retired from his law firm and wants his wife, Carrie, to take up golf and go on cruises with him. Carrie, 55, is balking. The interior design firm she founded is thriving, and she has no plans to give it up. In fact, she feels that at last she has found her stride and is fulfilling her dream of having a career. "We had the kids you wanted," charges John. "Now I'm retired, and all you can do is talk about your work. Who am I going to play golf and have lunch with?"

Sarah, 62, has taken early retirement and wants to travel, take university courses, and learn ballroom dancing. Her husband, Craig, 67, never intends to stop working. He wants Sarah to come and help him run his new and expanding import-export firm and is appalled by the idea of taking time to learn the fox-trot. "You can't be serious," he says with a snort. "You want to go off learning to dance and take workshops in bird watching. That is ridiculous. I am not even going to talk about that." He leaves and goes back to work.

If couples have a secure bond, they can work through such impasses and reach a compromise that is satisfactory, or at least tolerable, to both. When Carrie and John can have a calm and open conversation, in which they share their needs and fears and he can hear how deeply important it is to her to see her company grow, he actually offers to help her with drawing up contracts. She agrees to take at least six weeks every year to travel with John, and they plan their first trip together. She draws the line at golf, however! This couple's bond offered them a secure base on which they could stand

and explore each other's emotions, fears, and needs and find a way to deal with them together.

But Sarah and Craig are deadlocked. They cannot find a way to help each other deal with the shifts that make up this stage of their relationship. As Sarah becomes lonelier, without the companionship of her children, she gets angrier. As she demands more of Craig's time, he turns away more and works later. Sarah tells him in a high, agitated voice, "I have been married to you forever, but I can't remember a time when I felt you put me and our relationship first. I think our sex life and our kids held us together, but both of those are gone now. I am lonelier than I have ever been. I might as well leave. After all the hurts and injuries over the years, I don't know what I am doing here. I'd be better off by myself." Craig, who has no idea how to respond to his "hysterical" wife, turns away and closes down. "Fine," he mutters.

They will divorce. Indeed, retirement precipitates divorce in many long-term marriages. In Japan, where husbands routinely devote years to advancing their careers and rarely see their families, there is even a special name for this malaise—*shujin zaitaku sutoresu shoukougun,* which literally means "one's husband being at home stress syndrome," or "retired husband syndrome" for short.

Beyond Retirement

Retirement can be an acute crisis, usually lasting for a year or two at most. But then there are all those years after retirement. What keeps a love relationship going in our waning years? And how important is it?

A multitude of studies show that a positive, close relationship is one of the best predictors of longevity and physical and mental health. In one pioneering study, University of California psychologist Howard Friedman and his colleague Leslie Martin analyzed data on 1,500 middle-class folks who were born around 1910 in

California. The voluminous records traced their lives over eight decades until their deaths, detailing their experiences and habits through prosperity, the Great Depression, and two world wars. The notations included everything from the happiness of their parents' marriage to their career choice to the number of books they had in their home.

Friedman concluded that medical advances play a minor role in extending life span. "Most people who live to old age do not do so because they have beaten cancer, heart disease, depression, or diabetes," he says. "Instead, the long-lived avoid serious ailments altogether through a series of steps that rely on long-lasting, meaningful connections with others."

In other words, you can eat special organic and gluten-free foods, gulp down multivitamins, get yourself to the gym, and meditate into a stress-free zone, but the best tonic for staying healthy and happy into old age is probably toning up your relationship.

Being attached to a partner buffers us when illness strikes. Psychologists Anthony Mancini and George Bonanno of Columbia University questioned more than 1,500 elderly married couples living in Detroit. One member of each pair had a heightened sense of mortality due to a physical disability that made routine tasks, such as bathing, dressing, climbing stairs, and picking up heavy objects, difficult to accomplish on their own. The researchers discovered that self-esteem was higher and depression and anxiety lower in the handicapped people who had an emotionally responsive mate. Having a spouse who was willing to listen to their worries and made them feel loved made more of a difference to their mental well-being than did help with buttoning shirts, tying shoelaces, and the like.

Other studies reinforce these findings. It is emotional support—expressing concern and allowing partners to express their feelings—that sustains health and helps maintain optimum functioning of our body's cardiovascular, hormonal, and immune systems. And it is emotional support, not physical assistance or pragmatic advice, that most cushions us from the stress and strain of illness when it occurs.

Sybil, 68, suffers from chronic arthritis. She tells her husband, Harry, a spry 75, "You can help me best by being there for me and showing that I am important to you and you care. I don't need all that advice; it just distresses me. I need you." Harry looks puzzled, but I can assure him that she means this literally and that his closeness can make a real difference. As we grow older, secure connection with a loved and loving partner becomes an even more vital resource.

The end of this stage in a love relationship comes when one partner faces death. Even in this final transition, close ties can help both the dying and survivors, as a program at Princess Margaret Cancer Centre in Toronto affirms.

"I saw that partners were as devastated by the diagnosis of terminal cancer as the patient," observes psychologist Linda McLean, the program's director. "They were swept away on a wave of helplessness and anticipated loss as they watched their loved one decline. I knew that losing a partner put these people at risk for all kinds of health problems later." Indeed, survivors are physically as well as mentally very fragile in the year after a partner's death.

McLean and her team offered 42 couples who were dealing with a terminal disease at home either standard care—that is, practical advice on how to cope with the illness—or a modified version of my Emotionally Focused Therapy, which concentrates on shoring up the bond between spouses. Partners were counseled to talk about what they were facing, make joint decisions about how to control symptoms, and plan how to spend their remaining time together. In couples who received EFT, patients felt more heard, understood, accepted, and cared for by their spouses. And spouses felt less burdened by and more appreciated in their caregiving. They found satisfaction together in reviewing the dying partner's life and in creating a narrative of their time together.

"This project was so intense and so rewarding for everyone involved in it," says McLean. "It was a privilege to be with people in their last days and moments and to feel that we had helped them reach some kind of serenity and calm. After the patients died, the caregivers also expressed tremendous gratitude to us. They told us that the sense of resolution and connection they felt with their partner was invaluable. That this had helped with their helplessness and their grief."

Secure connection to a loved one not only helps us handle grief better and experience fewer traumatic symptoms but also can nurture and sustain us the rest of our lives. This is contrary to conventional wisdom, which urges survivors to let go of the departed so as to turn back to the living. We do need to accept that a partner is gone, but we can also hold on to our bond with our loved one by accessing memories or imagining exchanges and using them as a source of strength and comfort.

"It was so hard to lose him," a friend who lost her husband told me eighteen months after his death. "But now, when I see something lovely, like the snow falling very gently on a winter evening, I find myself telling him how beautiful it is, and knowing how much he would like it makes that moment even more beautiful. And when I am down, I remember how much he loved me. I still feel loved. So I am okay."

Her thoughts immediately brought to mind a line by Elizabeth Barrett Browning: "I shall but love thee better after death." Some pains are sweet.

Life is a series of transitions and transformations. One day you realize that you want to marry this man who last year was just a friend; then you turn around and you find you are having a fight on your seventh wedding anniversary. One day you run home with the news that you are pregnant; then suddenly your baby is an adolescent, and the next day he gets married. One day you and your husband walk into a new retirement condo, just the two of you. The next day you watch him as he picks up his tiny granddaughter. One day perhaps you sit and remember all the fights you had when you were newly married and feel amazement that you are still with, and still love, this person who always drops his socks on the floor and gets stupid in arguments.

If we are honest, what we hope for is that, in all these transformations, we find a way to hold and be held by our loved ones. Each shift at each new stage tests old ways of connecting and requires that we renew our bonds. This is what makes life worthwhile and what keeps us healthy and happy as we move inexorably from milestone to milestone.

EXPERIMENT 1

Think of a transition you have been through in a key relationship. Perhaps moving to a new apartment, renovating your house, leaving school, finding a job, or switching careers. Choose any period when you had to adjust to an uncertain situation.

How did this stress influence your day-to-day interactions with your partner?

See if you can write down one way each of you responded that invited the other person to draw closer so you could work as a team through the transition.

Now write down one way you made it harder for you and your partner to come together and help each other through that period. See if you can share these with each other.

EXPERIMENT 2

Here is an example of a couple learning to work together as their life shifts under their feet:

Linda tells her husband, Eric, "We are dealing with this whole move into retirement differently. We still have a mortgage, you know. I am working hard, selling my cosmetics line, and even though I told you it was okay for you to retire—after all, you'd been in the job for thirty years, and it was killing you—now I am resentful. You are not pulling your weight. My friends say that once the kids leave and work slows down, things get better. But I am taking care of my mom as well, and right now I just feel angry with you. I know I like my work, but there you are playing in the universe while I am . . ." [She throws up her hands, and tears well up.] "And then you just dismiss my worries. You get defensive, and that drives me wild."

Eric responds, "Look, I intended to work part-time on projects, but they just haven't come through. And I like being retired. At last I am not stressed out. I like having time to read and go to the gym. I like not being driven. And I kept us going for thirty years, so don't you tell me to feel bad about not working. We do have enough money, you know. We could easily downsize and not have a mortgage. But we can't seem to talk about this together. You get mad like this, and then I freak out and shut down. I guess I do get defensive. But we know this dance. We've done it for years. And you're right. I'm starting to avoid these conversations. I am starting to avoid you! And I know that doesn't work. But you're talking to me these days like I have turned into some kind of lazy loser. I want you to accept me a little here and trust that I don't want you to feel stressed and scared about money or caring for your mom or anything." [He laughs.] "Even though you never use that word—*scared*—I think that is what I hear when I calm down and look beyond your irritation. I am trying to listen to you. Tell me what you need here. We can work this out."

Linda calms down and admits she has a hard time articulating what she needs from Eric. This has always been hard for her. She agrees that she gets critical and admits that this isn't fair to him. They talk about how he can support her emotionally, and at the end of the conversation he has also agreed to go visit her mother a couple of times a week and to take a paying project that has just come up for the following weekend.

What did Eric do here that helped both of them keep their balance, stay connected, and problem-solve some of the issues in the transition they are going through? See if you can name at least two things he said that helped his wife.

EXPERIMENT 3

Think about a transition you are in right now in your personal or family life, or the next transition you will be in or hope to be in. It might be starting a family or going into retirement.

Write down the main way that you think your partner can help you through this transition. Be specific: What exactly could he or she do and when? Write out how you would ask for this help.

CHAPTER 7

Unraveling Bonds

Love never dies of a natural death. It dies because
we don't know how to replenish its source, it dies of
blindness and errors and betrayals. It dies of illness
and wounds, it dies of weariness, of witherings, of
tarnishings, but never of natural death.

— *Anaïs Nin*

My client Sam, a small and vociferous man who runs the
local deli and insists on bringing slices of extremely
smelly salami into my office for all the staff, is carrying
on again about how hard love is. "I am so sick of this," he mutters.
"It's always the same. My last relationship wasn't any different. You
go through all the falling in love bit and you get married. And everything is hunky-dory. For about a week. And then what? It all
starts to go wrong. My buddy Al says that that is just the way
women are. Never satisfied. What man really understands women?
One minute you are Mr. Wonderful, and the next, she is talking
about divorce and who will get the house. And who knows what
happened? You are the same guy. But suddenly it's all 'glass

179

empty.' I give up. Women are just too hard. And maybe we are just not meant to stay together forever anyway. My buddy says that it's just nature. We are supposed to move on. Or she is just with the wrong man. Mr. Right has gone missing here."

His wife, Marcy, reacts with a smile of such freezing contempt that I can feel the room icing up.

Sam is not done. He turns to me and slams the back of one hand down on the palm of the other. "Face it, psychologist lady, this love stuff just don't work, and no one knows why it goes from all huggy-wuggy to dust in your mouth in a moment. Isn't that a fact?"

I sit up in my chair. "Well, no, in fact, we now know so much about..." I begin, but then I realize he is way too hurt to hear me. I have seen this desperate bewilderment before.

Sam, or, rather, his buddy Al, has hit on all the old saws—and one new one—about why relationships fail. Sam should stop listening to his buddy. Al's all wet.

Let's examine Sam's and Al's assumptions one by one.

1. *The Alien Argument.* Men and women are just too different to ever get along, or, as John Gray so entertainingly put it, "Men are from Mars, women are from Venus."

Here's what we really know about sex differences. Men and women, in actuality, are remarkably alike. Really significant differences appear in only four areas. Three are cognitive: verbal facility, mathematical skill, and visual-spatial ability. Women win the first hands down—they use more words and express themselves better than do men. Men do better when it comes to working with numbers and calculations and being able to mentally manipulate two- and three-dimensional figures, but these abilities appear largely linked to expectations. If you tell women that tests of these skills are "gender neutral," they tend to perform as well as men.

Only one area of significant difference is psychological, and that is aggression. Men are quicker to anger and turn threatening or violent. In every other psychological aspect, the stereotypes fail. Adolescent girls are commonly held to be plagued with doubts about themselves, their attractiveness, and their talents, but boys have just as many self-esteem and self-confidence issues. Adult women are lauded as caring nurturers, but men are equally likely to be warmly supportive of offspring, family, and friends.

But surely women must be more empathetic, right? On a physiological level, there is no evidence for better mirror-neuron functioning in females. Psychologist William Ickes at the University of Texas conducts simple, real tests of empathy in which pairs of people sit and interact, then separately watch a tape of their exchange and report to the researcher what they felt or thought at particular moments. This tape is shown to the other person and stopped at these moments. The other person is invited to infer what his or her partner was feeling or thinking, and their guesses are checked for accuracy. Ickes concludes that men and women have the same basic ability. Differences only emerge when people are explicitly told that they are *expected* to act in a certain way because of their sex; then they try harder. Men who are told that women find nontraditional, empathetic males more desirable immediately improve their performance in this kind of task.

Of the four areas with sex differences, only two count in relationships: verbal facility and aggression. Women are more likely and better able to verbalize their feelings and needs than are men. They have more training: mothers talk to little girls in more elaborate ways about their emotions. And men, when they are anxious about their bond with a partner, are more likely to become physically antagonistic or to withdraw and evade. In day-to-day conflicts in love relationships, women tend to be more vocal demanders, while men tend to use silence to distance and defend. But even this difference tends to disappear when it is the man who wants a change in the relationship.

2. *The Soul Mate Claim.* This belief is perhaps the one most voiced by partners in distress. It has elements of the Alien Argument, but adds a personalized fillip. It goes something like this: "You're wildly emotional/incredibly controlling; I should have seen it but I didn't; you can't be fixed; I need a different type of person. You are not *the One*."

Online dating sites try to convince us that they will match us with our perfect partner, but we all know that our assumptions about what Mr. or Ms. Right looks like are irrelevant when it comes to who will capture our attention at a Saturday night party—let alone who will make us happy for life. Recent research finds that in face-to-face interactions, people are not particularly attracted to or romantically interested in folks who match their stated ideal-partner characteristics. Ideals and profiles are just lists of labels; in real encounters, factors like rapport and shared humor are more telling. In fact, I remember a stunningly handsome young man telling me on our first date, "I am not going to meet all your expectations"; I married him, of course.

Dating sites imply that the Perfect One is out there. One day someone is going to sue them for fraud. Personally, I like the propinquity theory about the way people choose their lover. Propinquity means nearness. The one who looks perfect is the one you are standing next to when your attachment system kicks in.

She happens to smile at the precise moment when you clue in to how alone you feel.

Of course, there is some truth to the idea that most of us gravitate toward mates who are similar to us and share the same values and interests, reflecting our implicit understanding that it's easier to get along with someone just like us. But despite the beliefs of our starry-eyed, romance-saturated society, there is no such thing as a perfect soul mate. Any partner we choose will hurt us at one time or another. No relationship, even the most ideal, has unwaveringly smooth sailing; there will always be squalls and storms that roil the waters. There will always be differences between lovers. How lovers allow their differences to affect the bond between them is the issue.

3. *Nature's Song Says Move Along.* Advanced by evolutionary biologists and taken up by the popular press, this is the newest explanation for why relationships fail. Love is a childish fairy tale. Evolution has programmed us to have short liaisons that last only until we're assured that our offspring have a strong chance to survive on their own. Then men, in particular, are meant to move on and spread their genes around so as to better ensure survival of the human race.

The trouble with this one is that it looks at a personal process — what happens to Dick and Jane — through the lens of ultimate causation, that is, in terms of an overarching universal principle about why a process exists in the first place. When a man flirts at a party, I don't believe that he, even on an unconscious level, is thinking about passing on his genes to the next generation. He may, though, be thinking about taking off his jeans for the lovely lady he's laughing with.

All three of these positions are defeatist and demoralizing.

There's no room here for accommodation, for improvement, for success. They're all doomsday scenarios.

Psychologists have also come up with theories about why relationships go off track. When I began training to work with couples, the most popular idea was that we all simply repeat with our lover the struggles we experienced in childhood with our most powerful parent. We project that parent's image onto our lover, went the theory, and act out old conflicts, actually manipulating our lover into acting out our scenarios and affirming our worst expectations. In a sense, our lover's actual responses are deemed irrelevant; our own personal neurotic need to repeat past patterns is seen as the key factor. This theory ignores the power of present interactions. Gradually, though, it gave way to clearer, simpler explanations of relationship derailment. Feminist scholars, for instance, have said that inequality is the downfall of relationships and that power struggles over tasks like sharing the housework are key.

As the systematic observation of couple interactions became more common, therapists became obsessed with two ideas: that conflict destroys love relationships and that distressed couples lack the skills to resolve such disputes. But as psychologist John Gottman, who has viewed many pairs in his famous Love Lab at the University of Washington, has pointed out, all couples fight, and happy couples really don't use the skills that are highlighted in traditional couple therapy. Among these are calling time-outs when fights get hot and taking turns speaking and repeating what the other just said (known as active listening). So how important can these skills be?

Before we gained love sense, it was hard to offer an incisive explanation for how love fails. Theories that concentrate on bad behavior in conflicts and lack of communication skills are focusing on the symptoms of couple distress rather than the root cause: the overwhelming fear of being emotionally abandoned, set adrift in the sea of life without safe harbor. It is that fear of emotional disengagement that precipitates the demands, criticism, arguments, and silences that mark troubled pairs. What we've missed for so long is that discord is almost always an unconscious protest against floating loose and an attempt to call, and even force, a partner back into emotional connection.

It's useful to look at the dissolution of love relationships in two ways: as a gradual erosion or unraveling of a bond over the course of many fights and silences or as an abrupt shattering of a link as the result of a traumatic injury or betrayal. Whether it is a slow wearing down of hope and affection or a sudden cataclysm that demolishes trust and commitment, it prompts a primal panic and the playing out of a survival script.

THE SLOW EROSION

John Bowlby's original understanding of relationship distress was framed around one word: *deprivation*. Looking at unhappy partners through the new lens of attachment, we see not only what is obviously corrosive in a relationship—that is, the turning against each other in conflict—but also what is missing. When love begins to erode, what is missing is attunement and the emotional responsiveness that goes with it. As responsiveness declines, partners become more vulnerable, and their need for emotional connection becomes more urgent.

The potential for conflict increases as partners are filled with unruly emotions that they do not understand, and find themselves out of sync with each other. Angry protests at the loss of connection escalate. The repair of specific hurts becomes more and more challenging. A slow unwinding of the tie begins. Lack of comfort and closeness feeds distrust and disagreement, and each failed attempt at reconnection and repair breeds more distance. As any sense of safe haven is lost, the old cliché that we build walls when we need bridges comes true.

When emotional starvation becomes the norm, and negative patterns of outraged criticism and obstinate defensiveness take over, our perspective changes. Our lover slowly begins to feel like an enemy; our most familiar friend turns into a stranger. Trust dies, and grief begins in earnest.

Annette, a lawyer in her early thirties, tells her husband, Bill, "I guess, when I look at it, all this didn't really start with the fights, did it? They were just the fallout. I just never grasped what was happening. I was so into building my career and kind of growing up, becoming successful and being a mom to our son. I guess now, as I try to hear you through all my frustration, I know that I did push us to the side. I was so caught up in things to do. Running faster and faster. I just didn't want to hear that you felt left behind or unimportant. I didn't listen, and when you got mad, I dismissed it as part of some midlife crisis of yours. I didn't want to fight. I thought the best thing was to kind of calm you down and trust that next morning it would be all right. I thought the fights were the problem... if they just stopped, then... But then there was this wall, and you were gone somehow. Now you've stopped turning to me. I guess maybe this is what they call falling out of love. Is it?"

Bill turns to her and says quietly, "I just gave up, Annette. You weren't there. I gave up. I couldn't stand the empty spaces between us anymore. I couldn't stand there, feeling naked, and ask—and have you tell me one more time to wait till you were less busy." Her face crumples.

Research confirms that erosion of a bond begins with the absence of emotional support. Psychologists Lauri Pasch and Thomas Bradbury of the University of California asked partners to solicit advice from one another about something they would like to change about themselves. They found that unsupportive behavior—minimizing the scope of the problem, discouraging the expression of feelings, offering offhand or unhelpful advice, insisting that their partner follow recommendations—was especially predictive of relationship distress. This result stood out, even when the effect of a partner's anger and contempt during fights was taken into account. Pasch and Bradbury conclude that the quality of positive support—reassurance that a partner is loved and esteemed and is capable of taking control of his or her life—is the most crucial factor in the health of any relationship.

This kind of result parallels the work of Ted Huston and his team, who questioned 168 couples at four different points in their marriages: at eight weeks, two years, three years, and fourteen years. Researchers went to each pair's home, asked them to fill out questionnaires, and interviewed the partners separately; nine follow-up telephone interviews were conducted over the next three weeks. Questions focused on positive and negative behaviors, such as how often a partner expressed affection or criticism. They analyzed this deluge of data, looking to see whether any particular behavior early in marriage is associated with its later stability. They found that the chief predictive factor in partners who split was not how often they fought as newlyweds but how much affection and emotional responsiveness they had shown each other. Couples who broke apart had been less demonstrative and responsive as just-marrieds than those who stayed together.

Huston concluded that it is not negativity per se that undermines partners' love for each other. Fights can be tolerated, provided there is support and affection in a relationship. Decreases in positive connection create "disillusionment" and precipitate distress. The absence of positive, intimate, supportive exchanges has been compared to a virus that takes down the body relational. Conflict is the inflammation that results from this virus; it is an attempt to solve the problem of lack of emotional responsiveness from a partner. In a troubled love relationship, problem solving and practical assistance alone will not be curative.

If erosion begins with loss of connection, the second stage is the escalation of conflict, especially negative patterns, such as demand-withdraw, that actively destroy any sense of emotional safety partners have with each other. I call this demand-withdraw two-step the Protest Polka; it is an objection to the separation and disconnection between partners. As both partners lose their emotional balance and attachment panic takes over, reactive rage and defensive numbing become more extreme and more compelling.

In their first appointment with me, George yells at Barbara, "I am a damned psychologist. I am supposed to understand this stuff, and I can't believe how angry I am. In my head, I rage at you as I am driving to work. I hear the sarcasm in my voice, and sometimes I wonder who I am turning into. I watch every minute for you to turn away from me. The more I push you to be with me, the more untouchable you become. But I can't stop doing it. I am married to the Ice Queen herself. You don't care about anyone but yourself. I want a wife, not a business partner."

Barbara crosses her long legs slowly, tilts her head, and calmly replies, "Then it would be good to try being polite and treat me like a wife. I do not see the point in all this shouting. And so you are right; I often prefer to leave and be elsewhere."

Lost in a dance they do not understand, their dance takes on a life of its own. George does not see how his anger makes Barbara fear that she is being rejected, and she does not hear the frantic call for connection underlying his irascibility. If they cannot find a way to step out of their polka and risk reaching for each other in another way, their bond cannot be repaired, and the end stage—disillusionment, despair, and detachment—will set in.

It helps if we understand the forces that are at work in these conflicts, in these struggles for love. Then we have a chance of

grasping the impact we have on each other. It helps if we understand the power of the two toxins in a love relationship—criticism
and stonewalling—and how they destroy emotional balance and
inflame insecurity.

Poisonous Criticism

"There is no such thing as constructive criticism," says John
Gottman. "All criticism is painful." He is correct. We never like to
hear that there is something "wrong" with us, or that something
needs changing, especially if this message is coming from the loved
one we most depend on. Psychologist Jill Hooley's work at Harvard
measures the impact of critical, hostile comments made by loved
ones and shows just how venomous disparagement by those we rely
on can be. This censure may even trigger relapse of mental illness,
such as depression.

Hooley's team looked at two groups of women—those who had
previously been depressed but had recovered and had been stable
for at least five months and those who had never been depressed.
These women were put in an fMRI machine and exposed to two
recorded scornful speeches made by their own mother about issues that had been raised in their relationship in the past. To me,
this sounds like Chinese water torture. I was shocked it got past
the ethics committee. The criticism sounded something like this:
"Your clothes are old and poorly fitting...Your newer things are
extreme in style in a way that isn't flattering to you. You need some
advice on style." But the researchers also made sure that the women
heard two speeches from their mom that contained praise, such as
"Stephanie, you have always had such a wonderful smile...This is
one of the things I have always loved about you, and I think others
do, too." So maybe it wasn't quite so bad.

A panel of judges rated the quality of the moms' praise and

criticism and agreed that there was no difference in the savagery of the criticism or the intensity of the praise that the two groups of daughters heard. But after listening to the critical speech, the women who had previously been depressed rated themselves more generally upset and described themselves in very negative terms, using such words as "irritable" and "ashamed." They also showed a smaller increase in positive mood after hearing praise as well.

But the question remained: Exactly how does criticism from an attachment figure affect neural responses in a way that prompts feelings and behaviors associated with depression? A powerful little sorting department in the brain called the DLPFC (for dorsolateral prefrontal cortex) is known to regulate the impact of external cues on the limbic system, the emotional brain. Depressed people consistently show decreased activity in the DLPFC; successful antidepressants boost activity. When the previously depressed women heard their mother's criticism, their DLPFC simply failed to activate, a finding that Hooley calls "striking." These women's brains were unable to switch into soothing and calming in the face of disparaging comments.

In earlier research, Hooley has found that patients hospitalized for depression have a two to three times greater risk of relapsing when they live with highly judgmental relatives instead of more approving ones. Hooley calls criticism from a loved one "low-grade punches to the brain." She also has found that censure by family members can be stressful enough to trigger relapse in people struggling not only from depression but also from schizophrenia and eating disorders. Criticism from loved ones rings the survival alarm bell in our brain; it sets off the deep-seated fear that we will be rejected and abandoned. It makes sense that such scorn makes it infinitely harder to hold on to our mental equilibrium and emotional balance.

No doubt, Mom is a powerful emotional figure in most people's lives. So are our romantic partners. In my couple therapy sessions over the years, I've noticed that partners usually have no clue as to the real impact of their negative judgments. When I first broach the idea that attacks overwhelm the partner on the receiving end with so much hurt and panic that he or she cannot deal with the disapproval and so withdraws and retreats, my clients often look at me incredulously.

"But mature adults should be able to deal with criticism. It's really just feedback," Carrie tells me.

"But this is *you* giving feedback that he is disappointing you, and you are your partner's main source of safe connection," I tell her. She still doesn't get it. I try again: "Even when things are going well with my own husband, if I hear disapproval and criticism in his voice, it's like a fire alarm. Anyone else's comments are more like a bicycle bell ringing. My brain tells me that keeping the approval of the one I rely on for a basic sense of belonging and safety is an urgent matter."

"You mean just because it's *me* and I have this special secure-making place in his life, my upset and blaming just freaks him out? It's alarming to him?" asks Carrie. I see her husband, Walt, nodding emphatically.

Criticism virtually guarantees that our partner will be caught up in fear and unable to hear our message and will become defensive and try to withdraw. Walt chimes in, "I just try to bat away your comments, but then you feel dismissed and you shout louder." Then he turns to me. "But if we work on strengthening our bond, then we will get to the place where we both feel so secure with each other that we can say anything and it won't ever sound like a fire alarm and we won't get all defensive, right?"

Wrong. When we love, we are always sensitive and vulnerable. But it is true that the more secure we are, the less we will get caught in negative patterns, such as demand-withdraw, that feed insecurity. Securely connected partners are also quicker to regain their emotional balance and bounce back from hurt and conflict than avoidant and anxious partners are. They are better at recognizing the impact they have on their lover and acknowledging that they have caused hurt. And they are better at repairing rifts, as you will see in the next chapter.

Toxic Stonewalling

We all use withdrawal at times when we are hurt or offended, or simply unsure and worried about saying the wrong thing. It is like a pause in the duet we do with our partner; it can allow us to gather our thoughts, find our balance. But withdrawal is toxic when it becomes the customary response to a partner's perceived blaming. And just as they do with criticism, my clients fail to recognize the impact this reactive distancing has.

"I don't understand why she is so angry with me," says Walt. "It can't be just because I go quiet. I space out because I can't deal with the hurt. I should man up and just shrug it off, but I can't. I get overwhelmed. Why can't she just wait till I recover a bit?" He then admits that, in fact, he never wants to resume the discussions because the emotions he feels are just too difficult for him to handle. I try to explain that a relationship is a dance. If there is a stumble, you pause to get your balance and then resume moving. But if you wait too long, your partner gets the feeling you don't want to continue the dance. She gets alarmed and angry and starts to protest. Conflict ensues.

But there is another level of withdrawal that is absolutely deadly in love relationships. This is when a partner turns to stone—still, silent, and completely inaccessible. This is a total negation of the bond. There is no engagement. It is one of the rules of attachment that any response is better than none. I must have heard the cry "I fight to get a reaction, any reaction" a thousand times. When we stonewall, the most extreme version of dismissal and nonresponsiveness, we mostly do so in order to cut off our emotions; we freeze and retreat into numbness. But when one dancer completely leaves the floor, the dance is no more. This catapults the remaining dancer into the terror of insignificance and abandonment.

The old adage "If you can't say something nice, it's best to say nothing at all," which was taught in manners class in my English school, is about the worst possible advice for a love relationship. The operative word here is *nothing,* and that is precisely what we leave a partner with when we routinely turn away, shut him or her out, and stop responding.

Stonewalling by a partner triggers an emotional meltdown that usually shows up as white-hot rage or intense grief. If we are not looking through an attachment lens, this extreme emotion looks bizarre; after all, the cue is almost a non-event. The other partner simply gave no response. Can simply doing nothing have such an impact? Looking at the prototypical attachment relationship can help us understand.

Psychologist Ed Tronick of the University of Massachusetts, Boston, demonstrated the effect of stonewalling many years ago in a series of landmark experiments with mothers and infants. The mother sits facing her baby, talking and playing with him. Then, on a signal from a researcher, she becomes silent and unmoving, and her face becomes flat and vacant. The infant typically picks up on her emotional absence very fast and starts trying to reengage her, opening his eyes wide and pointing and reaching. When the mother does not reconnect, the baby goes into high gear, shrieking for attention. When this makes no difference, he turns away from his mother, withdrawing from her. After a couple of minutes, he dissolves into frantic, panicked wailing. This wailing is difficult to watch; the infant's desperation is tangible. The researcher then signals the end of the experiment, and the mother then smiles and comforts her infant, who soon regains his equilibrium and returns to engaging with her happily. (You can find clips of the Still Face experiment on the Internet.)

I see the exact same sequence of events occur when an adult couple sits in my office. At some point, one partner shuts down and literally becomes still. Just like the infant in the experiment, the other partner will try to engage the still one, become insistent and aggressive, make attempts to turn away, but then, faced with no response and no relief from feeling abandoned, will finally dissolve into despair. In this most primordial of threat situations, our reactions and our responses are the same whether we are seven months old or fifty-seven years old. John Gottman and other researchers point out that male partners are more likely to stonewall than females. This may be because men are more easily flooded, less able to deal with strong attachment emotions, and slower to recover from stress. Some note, too, that men are more likely to be avoidant in style, and stonewalling is perhaps the ultimate avoidance strategy, short of leaving the relationship.

A partner's distress is magnified by the paradox of having his or her lover physically present but emotionally absent. The incongruity undermines any hope that effective action can be taken to reconnect. "I was never as lonely when I lived alone as I am living with Davida," Barry tells me. "I can't bear it. She is in the house, and it looks like I have a wife. We are a couple. But there is no connection. It is crazy-making. Disorienting. There is nothing I can do to get her to let me in. I am beyond frantic here." Chronic stonewalling, the refusal to engage, renders the other person helpless. The ultimate irony is that in trying to protect themselves, stonewallers imprison themselves. Virginia Woolf, in her book *A Room of One's Own,* put it perfectly: "I thought how unpleasant it is to be locked out; and I thought how it is worse perhaps to be locked in." There is no solution here to either partner's sense of isolation. There is no bond to count on.

Dead End

As the cycle of hostile criticism and stonewalling occurs more frequently, it becomes ingrained and defines the relationship. These episodes are so aversive and destructive that any positive moments and behaviors that occur are discounted and marginalized. And as a couple's behavior narrows, so do the partners' views of each other. They shrink in each other's eyes; the full panoply of their personalities shrivels down to a few noxious traits. She's a carping bitch; he's a withholding boor. In such a darkening environment, partners question every action or comment the other person makes. Psychologists refer to this as a process of escalating negative appraisal, where every response is seen in the worst possible light. Both partners become hypervigilant for any hint of slurs and slights, abandonment and rejection. They cannot give each other the benefit of the doubt, even for a moment.

This is what happens in my office when Zack looks up at a new picture on my wall while Helen is talking; she reads it as a sign of his "indifference and terminal arrogance" rather than simple curiosity about his surroundings. Helen snaps at Zack; he interprets this as a deliberate attempt to demolish him and prove his "incompetence" instead of considering that she might have had a bad day and is simply tired and discouraged. The way we perceive our partner and the meaning we attribute to his or her actions depends on our sense of emotional connection.

THE SUDDEN SNAP

For many couples, disengagement is a gradual process, sparked by a series of minor incidents and hurts that slowly swirls into a downward spiral. Another analogy is that of a pebble that lodges in a house's foundation, causing a tiny chink that over time widens until the edifice crumbles. For others, though, disconnection occurs abruptly, triggered by a single event—what we call a relationship injury or trauma. Then it's as if a bomb drops on the house, blasting out walls and shattering the foundation.

These events are cataclysmic, smashing a partner's sense of safety and leaving only pain and despair. Everything the injured partner assumed about the other—their relationship, their world—is overturned. Psychologist Judith Herman of Harvard Medical School calls injuries inflicted by attachment figures "violations of human connection." As with other traumas, a feeling of helplessness results. What is worse here is that the injury is caused by the very person who is the safe haven. This paradox leaves people confused and lost. They stumble around, unable to grasp what has happened or respond effectively.

Infidelity is the most obvious wound. "I cannot 'just get over' this," insists Ethan, addressing Lois, his wife of thirty years. "You ask me to put your affair aside, but every time you are late home, I wonder if you have found a new 'friend.' I can't seem to turn off the feeling that it could all happen again. I was blindsided last time; I never saw it coming. And I don't know how to get the love back. Even when it seems safe between us and you are trying to be loving, some part of me warns, 'Don't risk it. Don't ever let yourself be hurt like that again.'"

"I don't know how to heal this, either," says Lois dejectedly. "You're never going to forgive me no matter what I say or do." She turns away.

Ethan and Lois try to talk about her betrayal, but each time they do they focus on the wrong things and go down emotional dead ends. Ethan grills Lois for every detail of the affair, imagining that this will somehow give him back a sense of control. Questions such as "Where did you meet the last time you made love?" and "What did you do in bed?" can go on forever, and the hurt only grows. The partner who inflicted the injury often tries to dismiss its significance; this is always a mistake. Lois tells her husband, "Well, you had talked about how maybe we had grown in different directions, so I wasn't clear our relationship even mattered that much to you anymore." Ethan explodes. "Well, after all our years together, you could have damn well asked me, couldn't you!"

Couples like Lois and Ethan often are completely confused about how the affair happened, how destructive it has been, and how to deal with it. In fact, most people are not sure it is even possible to heal from affairs. It makes a real difference, however, if a couple has some understanding of love and the nature of the attachment bond they have violated.

First, unless a straying partner is extremely avoidant in terms of attachment (remember that avoidants are more open to one-night stands in general), most affairs are not primarily about sex; they are about the hunger for connection and not knowing how to satisfy this hunger with one's partner. Most times, an affair is an indication of a more profound problem. If you are dancing close and in tune with your partner, there is no place for a third dancer to enter. Often, the bond has begun to erode or failed to firm into a secure connection; frequently there have been preceding cycles of criticizing and distancing. But the partners have been unsure what that meant, let alone what to do about it. So they have been accepting the relationship as it is and accommodating to the lack of connection. Then suddenly the "My partner has turned to someone else and is having an affair" bomb detonates, and the relationship becomes obviously and overwhelmingly distressed.

Second, in terms of dealing with infidelity, it is the level and extent of the deception involved that seems to matter, rather than the nature of the sex acts themselves. It is the implications for attachment and trust that count. Christine tells her partner, "I can't do this. I hear you're sorry, and I even understand how it all happened. We have drifted apart over the years, and I did sideline you when you would get upset and try to talk about our relationship. I just didn't see it going anywhere. But the fact that you took this person to the cottage where you courted me and where we spent our *honeymoon* somehow makes it impossible for me to open up to you again. That you went there, among all the memories of us—the ghosts of us—and made love with someone else in our special place, and then *lied* to me about it for months and months, even when I asked you directly and showed how my doubts were driving me crazy. I can't get over that. I can forgive, perhaps, in time, but I cannot be with you, depend on you, without trust between us. And I don't think I can get that back."

Everyone knows that an affair can cripple a relationship. But other events may be just as momentous and damaging because they contravene our wired-in expectations that loved ones will be our shelter at moments of threat or distress. If we do not understand the incredible power of attachment and its impact on us, we can inadvertently hurt our partner deeply simply by not understanding what kind of response is required. All such disastrous events are marked by moments of intense need and vulnerability, when a loved one is called upon to provide responsive care and does not come through. In these incidents, the answer to the key attachment questions—"Are you there for me when I need you?" and "Will you put me first?"—is a resounding no.

In my clinical practice and research studies, I hear many tales of traumatic abandonment. The young wife, miscarrying and hysterical, whose husband froze up, unable to bring himself to touch or comfort her, and who called her brother to come and help. The immigrant who, missing her family, pleaded with her husband to pay for her sick mother to visit, and he told her to grow up and stop pining for what is past. The man who, after eye surgery, began to panic because his eye was hurting and entreated his wife to drive him to the hospital in the middle of the night, and she instead urged him to calm down. These failures of empathy and responsiveness create wounds that cannot be put aside or papered over. As with a break in a bone, they must be mended, or permanent disfigurement follows.

In my office Ken loses his temper and yells at his wife, Molly: "One hour after I lost my job, you were on the phone to your dad, persuading him to give me some position in his office. You never asked me how I felt or if I wanted that. You never talked to me or gave me any kind of comfort or reassurance. You just fixed the 'problem'; you assumed that I would take anything he offered me. You assumed I couldn't cope with this."

"This happened five years ago, and I am sick of hearing about it," Molly yells back. "You are just being immature about this. Most people would have seen what I did as supporting you." When we do not understand our own or our partner's pain, our attempts to address it often make it worse. Ken and Molly step into their usual angry protest–cool withdrawal pattern. The injury is then compounded. At the end of the conversation, Ken feels more alone and even less able to count on his wife.

Paul and Francine are locked on what happened the night her mother died, three months ago. "So I didn't instantly drive you to the hospital when you asked. You had been there all day, and the doctor said she was stable and you should go home and rest. You'd done enough for her. And then she got worse in the night, and you only got there just before she died. That is the crime. All the other things I have done over the years don't count." Francine weeps and tells him, "I begged you. But my feelings didn't matter. And you refused to help, even to take care of our son so I could find a way to go. Nothing I said made any difference. You didn't care that I needed to be there." A little later, she dries her eyes and tells him, "You make it all sound so reasonable, but you let me down, and I will never again ask you or show you that I need your caring. I dealt with all that pain on my own. But you tell me that this is not a problem. You still aren't listening." She gets up and leaves the room. Until it is healed, this wound will block her ability to turn to and depend on her partner.

Most people recognize these wounds on an instinctual, gut level when they are describing them, even if they have never heard of the new science of love, and many do not believe that they can be healed. But indeed they can, even when they occur in relationships that are already tottering; of course, the more secure the relationship, the more easily the wounds are healed. A decade ago, when my research team at the University of Ottawa first identified these kinds of events as attachment traumas, we realized that we didn't know much about how to actually help a person forgive and be open to the injuring partner again. The wisdom on forgiveness mostly came from philosophical texts, religious works, and moral tracts, which urged people to rise above anger and the desire for revenge. None offered a map for traveling from rage and grief to emotional resolution and renewed trust. So we set about charting the territory and developing a systematic method to promote healing. The Attachment Injury Resolution Model (AIRM) has proved itself effective in helping couples forgive and learn to trust again. (I'll describe this in detail in the next chapter under Healing Traumatic Injuries.)

A MOMENT-TO-MOMENT UNRAVELING

Let's look at one couple, Bonnie and Stan, and chart their steps to dead-end disconnection. First they slip into disconnection and deprivation and move on into recurring separation distress. Then, in a frantic attempt to reconnect, they go into the demand-and-withdraw dance; as one partner complains, the other stonewalls. Finally, hopelessness takes over, and they freeze up in despair, any vestiges of empathy and secure connection gone. I sometimes think of this as the three *E*s: erosion, escalation, and emptiness.

Bonnie and Stan started out with a storybook courtship. She was an award-winning teacher, he an ultrasuccessful lawyer. They met on the beach in San Francisco, and four months later, Stan proposed to Bonnie by the Golden Gate Bridge. They felt that they had found their home in each other, and their intimacy and lovemaking were more than either had ever dared to hope for. Already in their midthirties, they decided to have children right away. Bonnie quickly became pregnant, but her joy was tempered by a flare-up of a back problem, the result of an earlier skiing accident. The birth was difficult, and suddenly Bonnie was at home with a small baby and chronic back pain. Stan got promoted at this time and was soon working late every evening and most weekends. He knew Bonnie wanted to move to a new city, and for that they needed lots of money.

This is a classic setup for a relationship to go off track, for hopes to be dashed, and for connection to collapse into distance and despair. But what are the key moments when, if we could freeze them in time, we could see the bond between Bonnie and Stan beginning to weaken? I hear these moments as the partners tell their story and interact in my office.

• *The loosening begins with small moments of missed connection and a growing sense of deprivation.*

"Stan was never home during my pregnancy," says Bonnie. "It was like I was doing it alone." There may also be a galvanizing instant of disconnection and abandonment, often at a pivotal moment of need, that goes unrecognized and unaddressed. The answer to the central questions in any love relationship—"Are you there for me?" and "Will you be accessible and responsive to me?"—becomes "Maybe."

• *Little black weeds of doubt and distrust sprout.*

For Bonnie, they shot up when her baby was just eight months old and having a little difficulty digesting solid food. In pain from her back problem and sleep-deprived, she was now also worried about the baby. And she had no one to help out. Her sister, who had often come to her aid, had moved away, and Stan had just received a call from his mother saying that he needed to come home immediately because his father was falling into forgetfulness with dementia and she could not deal with it. Bonnie stood in the hall outside their bedroom as Stan packed his suitcase, and she pleaded with him to wait at least a few days. She told him she was completely overwhelmed and that he could not just leave her. But he did, and she vividly remembers the set of his jaw and the shape of his back as he walked away from her.

• *Moments of hurt and misattunement solidify into negative patterns.*

When Bonnie tries to talk about this incident, Stan is evasive. Now there are weeds growing everywhere in the relationship, stifling moments of shared pleasure and joy. "You refused to talk about this after you got home," complains Bonnie, turning to Stan. "It was like it was all no big deal for you. I told you how concerned I was about the baby, and you said she was fine. I told you how weary I was getting because of my painful back, and I'd ask you for a hug, but all you would do is give me long lectures about how my going to that energy doctor was stupid and a waste of money. When I said it was helping, you just laughed at me. But if your mother called and talked about your dad's health, you would be all kindness and talk with her forever. Your parents come first with you, then our kid, and then me—if you feel like it!" She goes on, sadly: "And if we talked about it, you just got defensive. The more we tried to fix it, the more upset I felt. I got more careful, more wary."

"No," says Stan. "You got angry, and suddenly no matter what I did, it was wrong." He ticks off the items on her list. "That doctor is a charlatan. I was working harder than I ever have so that we could make the move you wanted, and you were just throwing away the money. I was tired, too. I could have done with some hugs myself. You just don't appreciate the things I do. All you do is get irritated. I painted the baby's room when you were visiting your mother. Just to surprise you. Bought new pictures, got new fixtures, the ones you wanted—but did I get any recognition? No! Criticism is what I got." He stares at the floor and says, "I can't do anything right for her anymore, so I do less and less."

This couple's version of the Protest Polka, the demand-withdraw dance, now runs in an endless loop all by itself. The relationship is on automatic. The dial is on danger. They talk about trying to do some of the activities that brought them together, like going for a bike ride or to a concert, but they acknowledge that the tension between them seems to derail any pleasure they might have. They mention, too, that they seldom make love anymore. No surprise. Who wants to make love when you are teetering on the edge of an emotional abyss? Just as when a torn muscle fails to mend and constricts movement, so an injured relationship that isn't healed stiffens and becomes less elastic, spontaneous, and playful.

• *Even as a couple moves farther and farther apart, partners still make small bids for reconnection, but these now go unnoticed or are rebuffed.*

After a comment about her unhappiness, Bonnie makes a joke about how she can sound overly harsh sometimes. It is a subtle, even unconscious, attempt to change the tone of their interactions, the music between them. But Stan misses it. He isn't aware that she is reaching out to him. He is in protective mode and keeping his head down.

The pair's wariness makes them reject even obvious overtures to reconnect. "I do care about your back," Stan assures Bonnie. "I get scared and worried about you going to see that phony doctor. I want you to be healthy and happy." She looks at him sideways as he invites her to a new dance, but decides to play it safe. "Really?" she says. "I guess that is why you always go on about the money, then." He wilts in the face of her sarcasm and refusal to accept his concern. He tries again in a tired voice: "You basically don't believe I care for you, do you? You tell me I only care about my parents."

She stares at him. Every muscle in her face tells him he is right. Even when one person risks and reaches, the other does not see it or trust it or reciprocate. Bonnie told Stan at breakfast one day, "You don't kiss me anymore before you leave for work." Surprised, Stan replied, "Oh, well, right now my mouth is full of garlic sausage." Bonnie jibed as she left the room, "Right. We have our priorities right in this house." She feels she took a risk and got turned down. He feels slammed.

• *The couple's downward spiral gains momentum. Partners begin describing transgressions and each other in absolute terms.*

Stan finds himself muttering in the car on his way to work. "She always does that. Assumes the worst and doesn't give me a chance to explain. That is just who she is—angry and mean." He forgets that only a few months earlier, he would have been inclined to say,

"Her back was hurting her this morning. She wasn't trying to be mean. She'll be different when I get home."

Partners create a story of the relationship that fits their own personal unhappiness and centers around the faults of the other. The spouse first becomes a stranger who does not understand, then an enemy who inflicts hurt, and ultimately a devil who has engineered the Fall from Connection. This transformation of a partner from friend to fiend is helped along by memories of past negative attachment figures who suddenly spring to mind. Bonnie sees her volatile, alcoholic father in Stan and defines him as cold and uncaring, while he sees a resemblance between Bonnie and his first girlfriend, who once humiliated him by listing his deficits as a lover in front of his friends.

As both people become more guarded and share less, their respective "how we got into this mess" story expands. Their conversations are full of mutual attack, vindictive blaming, and stonewalling retreat. Anguish escalates, and safety dwindles further. They are so busy hitting back the other's hostile comments that they completely lose touch with the impact they have on each other and the relationship. Empathy, the ability to stand in each other's shoes, has vanished.

• *A sense of helplessness takes over. First one partner and then the other closes off and turns away to other activities and relationships where there is a feeling of competence and control.*

Stan no longer bothers to defend himself when Bonnie complains. He simply turns his back and walks away. He has a built a wall around himself that she cannot get through. At first, this agitated and frightened her. She tried breaking through by escalating her complaints and criticisms, but when that didn't work, she became resigned. Bonnie began spending ever-increasing time at her mother's home, while Stan built a studio where he can do some woodworking. They were discussing getting a divorce when they first came to see me.

⁓

Unlocking the key elements in the drama of relationship distress—the moments of disconnection, cycles of distorted signals, and sudden injuries that destroy our emotional lives and our families—is one of the big breakthroughs of the new science of love. This understanding marks the first step in our learning how to shape love. What you understand you can maintain, repair, and even enhance. We have come a long way in comprehending why and how we fight with and wall off the people we love most in the world. We now grasp why we move to protect ourselves from our most cherished ones as if they are our mortal enemies, and how this protection becomes a prison.

For Sam, the husband whose outpourings began this chapter, relationships were a mysterious force over which he had no control. In my office four months later, he gives another speech—a different one. Sitting back in his chair, he begins in a soft voice, "I'm feeling much better about us." He pauses to smile. "And about me! Seems like I can manage this closeness stuff after all. We still fight sometimes, but I don't feel like I am standing on the edge of a cliff anymore. I didn't know how to listen to her, so I just tried to push her into loving me. Feels like we have learned so much. We know how to hold on to each other now." Then he looks at me with what looks like grudging respect. "Maybe you guys know something after all, psychologist lady," he says.

I admit I cannot resist the urge to feel a little smug at this point. But we need to find out more about how Sam and his partner not only helped each other move out of distance and conflict but also learned to create a new loving responsiveness that made a haven out of a relationship that had become a hell. We need to look at what we know about renewing bonds.

EXPERIMENT 1

1. Sit quietly for a moment with a pen and paper in front of you. Then think about a typical day in your relationship. On a scale of 1 to 10, where 1 is not critical at all, 5 is moderately critical, and 10 is highly critical, answer the following questions:

How critical and disapproving is your partner of you? How critical and disapproving are you of your partner?

Write your answers down on the piece of paper. Don't worry too much about accuracy. You may decide on a particular day that your partner is very disapproving of you and give him or her a 10. This does not mean that your lover is a creep and you should drop the relationship. It could just be a bad day. What's important in this experiment is discerning the impact you and your partner have on each other.

Share your answers with your partner. Try to do this in the spirit of learning how you both experience your dance. See what you can discover. Of course, you can use this exercise as a chance to be critical if you wish, or you can use it as an opportunity to explore the effect of criticism on your bond.

2. Now sit quietly and reflect on any moments in the day when you feel the most trust for and safe connection with your partner. It might be when she calls your name as she comes in the door from work. It might be when you kiss goodnight. See if you can guess when your partner might feel

this way, and then ask if you are on target. If such moments are hard to find in your current relationship, or if you are not with a partner, see if you can recall such moments in a past romantic relationship, or even with a parent.

What is the message you get from your partner's response in these moments?

Do you and your partner have ways of turning down the heat during a fight and making it more likely you will end up connected?

Are there ways your partner could help you calm down and regain your emotional balance when you are both reeling and not hearing each other?

Angela tells her husband, "What helps a lot when I am getting scared is when you just turn and say, 'We can work it out, honey,' and touch my arm. Then even if we go back to fighting, somehow it's better."

EXPERIMENT 2

Below are three scenarios of disconnection and attempted repair. See which one most reminds you of your usual responses and strategies when you and your partner (or a past partner or another attachment figure) are out of sync and stepping on each other's toes.

1. Ed knows that he and Lily are not doing well. He especially notices that Lily is not touching him or making eye contact. He decides to ignore it and let sleeping dogs lie so that he won't spark a conflict. He hopes that things will just get better with time. He thinks that talking about it will probably make it worse.

2. Joel realizes that he and Alison have not made love in weeks. This alarms him, and he worries that she doesn't desire

him anymore. He brings up the subject and points out that they should make love tonight. He also says he wants them to make love three times a week from now on. He can't resist adding that he thinks she should be more flirtatious with him, even though she is a cold kind of person.

3. Rick wants Ina to tell him that his worries about the row they had last week are unfounded. He decides to tell her that he is really concerned that she is still angry with him and is unsure about how she really feels about him. He shares this and asks her to be open with him and, if she can, reassure him that everything is now okay between them.

See if you can imagine, and describe in a sentence, the likely impact of each strategy on the other partner. What would he or she do? Turn toward, away from, or against the speaker? How might the conversation evolve from there?

You could describe the strategies as withdraw and avoid, approach and demand, and approach and share. Write a likely ending for each scenario.

EXPERIMENT 3

Think of one positive love relationship you have now or have had in the past. How did you try to repair moments of disconnection and change the emotional climate of the relationship?

Successful couples tend to openly acknowledge moments of distance and their impact. They share their emotions and make clear statements about feeling hurt or regretting that they caused hurt. They frequently use humor and also touch while talking.

When your partner tried to repair and reconnect after a time of distance, hurt, or conflict, did you accept these attempts? Or did you reject them?

CHAPTER 8

Renewing Bonds

We said we'd walk together baby come what may
That come the twilight should we lose our way
If as we're walkin' a hand should slip free
I'll wait for you
And should I fall behind
Wait for me

—*Bruce Springsteen*

We know the moments when we find connection again:
the universe lights up. These instances leap from the
pages of novels, burn in our brains when we watch
them in movies—or even in dusty research tapes—and, of course,
entrance us when they happen in our own precious relationships.
Everything comes together: suddenly all the blocks roll away, and
there is an open, easy connection. But how did we get there? If we
don't know the path, how can we get there again?

Patrick, 45, a results-oriented businessman who unearthed my
now fossilized PhD thesis before our first session, tells me, "Look,
I am here because my relationship is on fire. Anna and I have been
together ten years. Two years ago we moved back here from Cal-

ifornia, just like she wanted, and I sold my company. But now everything is burning up. My wife is either spitting rage or ignoring me. I can't handle it anymore." He impatiently brushes a tear from his eye. "So just tell me what to say in this new kind of conversation that your research says can change a relationship, and I will say it. Then we can get done here in two sessions."

I see that he wants the pain to stop. I explain that first he and his wife need to be able to help each other out of the stuck conversations that fuel constant hurt and fear. They must create a secure base to stand on and get their balance before taking the risky steps involved in new, connected kinds of conversations. He is not convinced. He asks, "But aren't there some people who just do this new kind of talk naturally?" I agree, but add, "Some of us have been lucky enough to have had great relationships in which someone walked through this kind of conversation with us before, and we learned what it felt like." As John Bowlby said way back, "We do as we have been done by." I try to explain to Patrick that when we think of others as basically safe and caring, we tend to have seen and practiced constructive ways to handle our emotions and respond to a partner in the past. So we have more options stored in our memory.

Still, no one can be open and responsive all the time. You always need help from your partner. I tell him, "Renewing your connection is something you do together. You both help each other keep your emotional balance and turn toward each other and tune in. It's a dance. It's not something you can 'fix' all by yourself from the outside by just saying the right words." The final straw for him is when I add, "Many of us aren't comfortable even talking about our softer, deeper feelings and can't imagine sharing them with our partner."

He shakes his head, blows through his nose, and stares at the door. Then his face falls, and he whispers, "I don't know about feelings. I just know that I don't want to lose my wife and my family. My sons are four and six. I love them so much." I ask, "But this all seems like foreign territory to you?" He nods. Anna, a former high school teacher, stares at me with beautiful green eyes and says quietly and slowly, "We have never really talked about deeper feelings. We have never been to that place." We begin.

Ten weeks later Patrick and Anna have learned to recognize and curtail their version of the Protest Polka. In their dance, which they have named "the Maze," Patrick reacts to Anna's upsets with the children or her problem with insomnia with cool rationality. Both then twist around and around, but the more they try to find a way out of their mutual frustration, the more lost and confused they become. Patrick tries desperately to offer solution after solution while totally bypassing Anna's emotional distress. In response, she becomes even more distraught and berates him for being "unfeeling"; he counters by calling her "hysterical." "Now I get that Anna sees me as distant and unapproachable," says Patrick. "I stay away from feeling stuff and go into fix-it mode, and then she feels alone and hurt. I get that." Anna observes, "I don't think I understood my anger, either. I'd call out for him and get this office-manager response and a list of solutions like 'Stop being so sensitive.' So I'd just start up the accusations. But I've started to see how this hurts him. He does care. Maybe he just doesn't know what to do when I call."

Now that they understand their disconnection dance, they are breaking out of it quickly. Says Anna, "The other night, Patrick said, 'Hey, this is the "round and round but always lost" thing again taking over. I'm in fix-it mode. It's so easy to go there when I hear that you are upset and disappointed with us. So you must be getting that left-alone feeling right now, like I am immune to your hurt and don't care.' I was blown away when he said that. I just went and hugged him, and he made a little joke. [She scrunches up her nose.] It wasn't funny, but that was okay." She laughs.

But healing a relationship isn't just about recognizing and stopping destructive behavior. That is just the first step. The second, and even harder, step is actively working together to build a stronger and more durable emotional union. That requires dumping old notions—for example, that love operates in a fixed, steady state—and becoming more proactive, such as by being alert to the small rents in the fabric of emotional connection and knowing how to repair them. The process of renewing bonds, we have learned, is continuous and inspiring, taking emotional connection to a whole new level. It makes us more emotionally accessible, responsive, and engaged, and thus it leads to deeper bonding and greater relationship stability and satisfaction. It also transforms us as people. As we take risks and confront our vulnerabilities, our trust grows—not just in our partners but also in ourselves.

THE RHYTHM OF DISCONNECTION AND RECONNECTION

A love relationship is never static; it ebbs and flows. If we want love to last, we have to grasp this fact and get used to paying attention to and readjusting our level of emotional engagement.

"I just assumed that once you are married, you both know you are partners and you can kind of relax and take the relationship for granted," Jeremy tells Harriet. "You can focus on the big picture. You know I love you. We aren't mean to each other. I haven't been unfaithful to you or anything like that. Can't you just roll with the less romantic, less touchy-feely times?" Harriet sits up straight in her chair and declares, "No, Jeremy. I can't. Not anymore."

"Well, that is just very immature, then," Jeremy replies.

He is right in a way. In a good relationship, where we feel basically secure, we can fill in the blanks left by our partner's occasional emotional absence. We can substitute positive feelings from past encounters and accept that there may be legitimate reasons for the inattention. But only some of the time, and only if we know we can reconnect if we really need to.

Loving is a process that constantly moves from harmony to disharmony, from mutual attunement and responsiveness to mis-attunement and disconnection—and back again. But to really understand what happens, we have to zoom down into these inter-actions and atomize them. Remember Seurat's paintings: when we move in really close, we realize that the vast scenes are composed of thousands and thousands of little dots. Researchers are doing the same with love relationships. By freeze-framing videos of romantic partners talking or arguing, and of babies playing with a parent, they are discovering how love, without our being aware, is shaped, for better or worse, in micromoments and micromoves of connec-tion and disconnection.

Up close, this is what love looks like: I look at you with my eyes wide open, trying to capture your glance, and you catch my expres-sion, widen your eyes, and take my arm. Alternatively, you ignore my bid for your attention, continue talking about your thoughts, and I turn away. In the next step, we resynchronize and reconnect. I turn back to you and lean forward and touch your arm; this time, you get my cue and turn toward me, smile, and ask me how I am. This tiny, fleeting moment of repair brings a rush of positive emo-tion. Moments of meeting are mutually delightful. (I always think that if we stopped and verbalized our innermost thoughts at this point, we would say something like "Oh, there you are" or even "Ah, here we are together.")

It's important to emphasize that misattunement is *not* a sign of lack of love or commitment. It is inevitable and normal; in fact, it is startlingly common. Ed Tronick of Harvard Medical School, who

214

has spent years absorbed in monitoring the interactions between mother and child, finds that even happily bonded mothers and infants miss each other's signals fully 70 percent of the time. Adults miss their partner's cues most of the time, too! We all send unclear signals and misread cues. We become distracted, we suddenly shift our level of emotional intensity and leave our partner behind, or we simply overload each other with too many signals and messages. Only in the movies does one poignant gaze predictably follow another and one small touch always elicit an exquisitely timed gesture in return. We are sorely mistaken if we believe that love is about always being in tune.

What matters is if we can repair tiny moments of misattunement and come back into harmony. Bonding is an eternal process of renewal. Relationship stability depends not on healing huge rifts but on mending the constant small tears. Indeed, says John Gottman of the University of Washington, what distinguishes master couples, the term he gives successful pairs, is not the ability to avoid fights but the ability to repair routine disconnections.

We learn about mini-misattunement and repair in our earliest interactions. Tronick and his team have detailed what happens by analyzing videos of infants and their mothers playing a game of peekaboo that grows gradually more intense. At first the infant is happy, but as the game builds, he becomes overstimulated and turns away and sucks his thumb. Mom, intent on playing, misses this cue, and loudly cries "Boo" again. The baby looks down with no expression. He shuts down to avoid her signals, which are suddenly too fast and too strong for him.

There are two basic scenarios for what happens next, one positive, the other negative. In the first, Mom picks up the cue that her child is overwhelmed, and she goes quiet. She tunes in to his emotional expression. She waits until he looks up and smiles at him very slowly, and then more invitingly, lifting her eyebrows

215

and opening her eyes. Then she starts the game again. Misattunement and momentary disconnection shift to renewed attunement and easy synchrony. All it takes is a smile or tender touch.

In the second scenario, Mom ignores or doesn't get her baby's signal. She moves in faster and closer, insisting her child stay engaged with her. He continues to turn away, and the mother reaches out and pushes his face back toward her. The infant closes his eyes and erupts in agitated wails. The mother, annoyed, now turns away. This is misattunement with no repair, what Tronick calls "interactive failure." Both mother and infant feel disconnected and emotionally upset.

Over time, thousands of these micromoves accumulate until they coalesce into a pattern typical of secure or insecure bonding. Tronick notes that at just seven months of age, infants with the most positive, attuned mothers express the most joy and positive emotion, while those with the most disengaged moms show the greatest amount of crying and other protest behaviors. Those with the most intrusive moms look away the most. We learn in these earliest exchanges with our loved ones whether people are likely to respond to our cues and just how correctable moments of misattunement are.

Those of us who wind up securely attached have learned that momentary disconnection is tolerable rather than catastrophic and that another person will be there to help us regain our emotional balance and reconnect. Those who become anxiously attached have been taught a different lesson: that we cannot rely on another person to respond and reconnect, and so momentary disconnection is always potentially calamitous. Those who become avoidant have absorbed a still harsher lesson: that no one will come when needed no matter what we do, so it's better not to bother trying to connect at all.

We carry these lessons forward into adulthood, where they color our romantic relationships. "The past is never dead," wrote novelist William Faulkner. "It's not even past." Psychologist Jessica Salvatore, along with her colleagues at the University of Minnesota, studied the romantic relationships of 73 young adult men and women. They had all been enrolled since birth in a longitudinal study of attachment, and their relationship with their mother had been assessed when they were between twelve and eighteen months old. They were invited to the lab with their romantic partner, where they were interviewed separately. Then they were instructed to discuss a key conflict between them for ten minutes and then talk about areas where they were in agreement for another four "cool down" minutes.

Researchers videotaped these talks and observed how well the 73 adults could let go of their conflict and shift out of a negative emotional tone. Some made the switch quickly and easily; others persisted in talking about the conflict and brought up new issues; still others refused to talk at all. Those who were good at cool down were generally happier in their relationship, and so was their partner. And, as we might expect, those who had been rated securely attached as babies generally moved out of the conflict discussion most successfully.

But is a person's own attachment history the key predictor of stability in a romantic relationship? Or is a partner's ability to resolve conflict also a major factor? Salvatore assessed the 73 subjects two years later and found that even among those who had histories marked by insecurity, their romantic relationship was more likely to have endured if their partner was able to recover well from an argument and help them transition into a positive conversation.

I call this the buffer, balance, bounce effect. A more secure partner buffers your fears and helps you regain your emotional balance so you can reconnect. Then together, you both bounce back from separation distress, distance, and conflict. We are never so secure that we do not need our partner's help in readjusting the emotional music in our attachment dance. Relationship distress and repair are always a two-person affair; a dance is never defined by just one person.

Some of us, however, need more structured help in finding our way back to emotional harmony. Drawing from my discoveries in thirty years of practice and research and the findings of the new science of love outlined in these pages, I and my colleagues have created a powerful model for repairing relationship bonds, Emotionally Focused Therapy. The only intervention based on attachment, EFT is redefining the field of couple therapy and education. Sixteen studies now validate its success. Couples who have had EFT show overall increased satisfaction with their relationships and in the elements of secure attachment, including intimacy, trust, and forgiveness. Moreover, the more secure emotional bond remains stable years after therapy.

One of our newest and most exciting studies, discussed in Chapter 3, demonstrates through fMRI brain scans that after couples go through EFT and become more secure, holding the hand of their partner actually dampens fear and the pain of an electric shock. Just as predicted by attachment science, contact with a loving, responsive partner is a powerful buffer against danger and threat. When we change our love relationships, we change our brains and change our world.

The science of love allows us to hone our interventions—to be on target and aim high. The goal is to create lasting lifelong bonds that offer safe-haven security to both partners. Recently we have also created a group educational program based on my earlier book *Hold Me Tight: Seven Conversations for a Lifetime of Love* that helps couples take all we have learned in decades of research and use it in their own relationship.

REPAIRING BONDS MOMENT TO MOMENT

As we discussed in Chapter 2, happy, lasting bonds are all about emotional responsiveness. The core attachment question—"Are you there for me?"—requires a "yes" in response. A secure bond has three basic elements:

- accessibility—you give me your attention and are emotionally open to what I am saying;
- responsiveness—you accept my needs and fears and offer comfort and caring; and
- engagement—you are emotionally present, absorbed, and involved with me.

When these elements are missing and alienation and disconnection take over, renewing a bond that is truly coming undone is essentially a two-step process. First, partners have to help each other slow down and contain the circular dance that keeps them emotionally off balance and hypervigilant for signs of threat or loss. Relationships begin to improve when partners can stop these runaway cycles that create emotional starvation and attachment panic.

To curb these demand-withdraw cycles, we first need to recognize that they *are* cycles. We get caught up in focusing on our partner's actions and forget that we are players, too. We have to realize that we are in a feedback loop that we both contribute to. When we see that this is a dance we do together, we can stop our automatic, blaming, "You always step on my foot" response. This allows us to see the power and momentum of the dance and how we are both controlled and freaked out by it.

Prue accuses Larry of being hypercritical. "He's always complaining about whatever I do—how I cook, how I make love. I feel picked on all the time. It's devastating." Larry argues that Prue al-

219

ways refuses to talk seriously about any problems they're having. "She just goes distant. I can't find her," he says. In our sessions, they've now realized that they are prisoners of a pattern they call "the Pit." I encourage clients to give a name to their pattern to help them see it and begin to recognize that the pattern, not the partner, is the enemy. They have both unwittingly created this enemy that is taking over their relationship, and they must work together to wrest their relationship from its clutches.

Now we can explore the triggers and emotions that shape the pattern. Prue and Larry recount a specific incident when they fell into the Pit, and we bring it into high focus and play it in slow motion, scrutinizing each detail, until its impact on each partner and their bond is clear. They were on holiday in Europe after a period when Prue had been away taking care of her dying aunt and Larry had resented her absence. They were in a station heading to catch a train when Larry suddenly realized that it had begun moving. Afraid they would miss it, he jumped on the step and yelled to Prue, who was carrying a coffee cup, "Run." Larry shouted to the conductor to slow down and held his hand out to Prue, but she froze. Finally, she grasped his hand and struggled onto the train, out of breath. Larry turned to her and said, "You are so damn slow." Shocked and hurt, she refused to speak to him the rest of the journey. Inside, she vacillated between rage at Larry's reprimands and dread that she really is too "slow" and too flawed for him to love. She shut him out and, preoccupied with her own fears of inadequacy, began a downward spiral into depression.

I turn to Larry and we go over and over this incident moment by moment and tune in to the emotions he was feeling then and how they reflect his overall feeling about Prue and their relationship. He says he feels "agitated" when she does not keep up with him on hikes. He notes she doesn't take her arthritis medication consistently. "I get anxious when she does not stay with me. I can't

220

count on her." He recalls the image of "distance" that flooded him when the train started to move off and Prue froze. "She wasn't running, working to be with me," he says. He felt panicked. Larry then begins to talk about his sense of isolation when Prue stayed with her aunt for three months and his habit of dismissing, or "pushing down," this frequent feeling. Sometimes he can't, though, and it rises up and engulfs him, and he winds up being angry and sarcastic. He begins weeping as he realizes just how much he needs her and is afraid that she will remain "unavailable." The slide into the Pit begins with attachment terror.

For Prue, too, the terror that freezes her and turns her away from Larry is a hopeless certainty that she is flawed and worthless, so rejection is certain. As they recognize and find their balance in these emotional moments, they can see the drama of distress as it occurs in their everyday life and then help each other halt its momentum. They can limit the extent of the rift between them and find a secure base. The next night, Larry lashes out, and Prue responds, "Is this a panic moment for you? I am not going to freeze up here, and I want you to slow down." Each partner begins to see the other in a new light: Prue sees Larry as afraid rather than judgmental and aggressive, and he sees her as protecting herself from rejection rather than simply abandoning him and "sulking."

Recent research by psychologist Shiri Cohen and her colleagues at Harvard Medical School confirms that partners do not suddenly have to become masters of empathy or emotional gymnasts in this kind of process. Partners, especially women, really respond to signs that their loved one is trying to tune in and actually cares about their feelings. This, in and of itself, creates a new safety zone where partners can begin to expand their dance steps and take risks with each other. New ways of dealing with emotion shape new steps in the dance, which in turn shape new chances for reattunement and repair. But this ability to keep miscues and missteps in check is not enough.

The second step in renewing bonds is much harder but more significant. This is when we move into powerful positive interactions and actually reach for each other. Specifically, withdrawn partners have to open up and engage on an emotional level, and blaming partners have to risk asking for what they need from a place of vulnerability. Partners have to tune in to the bonding channel and stay there. They find this process risky, but if they follow it through, their relationship becomes flooded with positive emotion and ascends to a whole new level. This process is not only a corrective move that kick-starts trust but also, for many, a *transforming* and *liberating* emotional experience.

These experiences are deeply emotional; partners each reach for the other in a simple and coherent way that pulls forth a tender, compassionate response. This begins a new positive bonding cycle, a reach-and-respond sequence that builds a mental model of relationships as a safe haven. It addresses each person's most basic needs for safety, connection, and comfort. These kinds of primal emotional moments are so significant that, as with all such "hot" moments, our brain seems to faithfully store them, filing them in our neural networks as the protocol for how to be close to others. Our follow-up studies of EFT couples show that their ability to stay with and shape these emotional moments is the best predictor of stable relationship repair and satisfaction years later.

So what actually happens in these exchanges—I call them Hold Me Tight conversations—when real connection begins to form and a couple moves from antagonism into harmony? Until recently we have not known what specific responses in intimate exchanges make for tender loving bonds between adults. We have had, to quote psychologists Linda Roberts and Danielle Greenberg of the University of Wisconsin, "a typology of conflict . . . but no road maps for positive intimate behavior." Years of watching couples reconnect in a therapy that deliberately builds bonds can offer us just this.

In Hold Me Tight conversations, couples have to handle a series of mini-tasks. Partners, whether pursuing and blaming or defending and withdrawing, attempt to:

• *Tune in to and stay with their own softer emotions and hold on to the hope of potential connection with the loved one.*

John: "I did snap at you. But when I look inside, it's that I find it worrying, upsetting that you go out to those clubs with your girl-friends. It somehow messes me up. It's hard to tell you this. I am not used to talking about this kind of stuff."

• *Regulate their emotions so they can look out at the other person with some openness and curiosity and show willingness to listen to incoming cues. They are not flooded or trying to shut down and stay numb.*

John: "I feel a little silly, kind of wide open saying this. But there it is. It doesn't work to deny it and say nothing. Then we get far-ther apart. Can you hear me? What do you think?" His wife, Kim, comes and hugs him.

• *Turn their emotions into clear, specific signals. Messages are not con-flicted or garbled. Clear communication flows from a clear inner sense of feared danger and longed-for safety.*

John: "I know I sometimes go off about you being tired after com-ing home late or the money you spend. But that is not it. Those are side issues. It reminds me of past relationships. I guess I am really sensitive here. I really find it difficult. It scares me. I wanted to run after you and say, 'Don't go.' It's like you are choosing them and the club scene over me, over us. That is how it feels." His eyes widen, showing how anxious he is.

• *Tolerate fears of the other's response enough to stay engaged and give the other a chance to respond.*

John: "You aren't saying anything. Are you mad now? I want us to talk about this kind of stuff when I get unsure of us and not push things under the rug. I want to hear how you feel right now." Kim tells him she is confused because she feels loyal to her friends but that his feelings are important.

• *Explicitly state needs. To do this they have to recognize and accept their attachment needs.*

John: "I want to know you are committed to us, to me. I want to feel like you are my partner and that nothing is more important than that. I need that reassurance that my needs matter. Then I can keep taking risks here. I am out on a limb otherwise."

• *Hear and accept the needs of the other. Respond to these needs with empathy and honesty.*

John: "I know I have been kind of controlling in the past. It's a bit hard to hear you talk about it, but I know you need to make choices, and you have fun with your friends. I am not giving orders here. I want to know if we can work this out together."

• *React to the other's response, even if it is not what is hoped for, in a way that is relatively balanced and, especially if it is what is hoped for, with increased trust and positive emotion.*

John: "Well, you have tickets for the concert, so I guess you will go. I can handle that. I hadn't really shared with you openly about this. It helps if I feel included somehow, if you tell me about it afterward. And I appreciate that you are listening and telling me that you can consider how I feel about this." Kim tells him she still feels scared to put herself in his hands completely. Her nights out are her statement that she is still holding on to her boundaries and showing she can stand up to him. But she hears his fears. She tells him that she does not flirt or drink too much on her outings, and she reminds him that she is going out less often now.

• *Explore and take into account the partner's reality and make sense of, rather than dismiss, his or her response.*

John: "I don't want to tell you what to do. I know this upsets you. You have good reasons for this. I get that you are not trying to hurt me. I don't want you to feel dictated to. I just get anxious about this stuff." He reaches out for her, and she turns to him and holds him.

When this conversation goes off track, John—and hopefully Kim—can bring it back and stay with the main emotional message, the need to connect. For example, if John gets caught up ranting about the "seedy" clubs she visits, she is able to stay calm and soothe him by telling him that she is concerned that he worries about this, and this brings him back to talking about his fears. Both partners help each other keep their emotional balance and stay in the deeper emotion and bonding channel. John is attempting to repair his sense of disconnection, and he does it by exploring his own emotions and engaging with Kim. In the past he had tried criticizing his lover's taste in friends or making deals about how many times each could go out without the other every month. Now he goes to the core dialogue in an attachment relationship, the one that matters most, where the question "Are you there for me?" is palpable. He shares and asks for her emotional support, for her help in dealing with his attachment fears.

This is very different from the way attempts at connection show up in distressed relationships and even in routine interactions in relatively happy relationships. We often bypass the attachment emotions and messages. We do not say what we need. Our signals to our loved one remain hidden, general, and ambiguous. Hal tells Lulu, "I don't think I have ever asked you for affection. It's not what I do. When you just give it, everything is fine. But when you get depressed...So then I say, 'Want to watch a movie?' or 'You should go for a walk and cheer up.' But you turn away, and in two seconds flat I am enraged. In my head, I am still thinking it's about the movie or you not taking care of yourself. Not that you have gone missing on me." When Hal can express his sense of loss at Lulu's withdrawal, they can deal with it and her bout of depression differently—that is, in a way that leaves them more connected rather than less.

The most intense and attachment-focused Hold Me Tight conversations build tangible safety and connection, even in secure, happy relationships. They can occur at times when partners do not feel disconnected but simply want more intense intimacy. Lulu opens up one night and tells Hal of a moment after their lovemaking when she felt herself "sinking into a certain soft place where we just belong and belong and there is no more fear of risking." He responds and shares his similar feelings. Each time these lovers share their "soft places" and their need for each other and respond with empathy and care, they offer their loved one reassurance that he or she is the chosen, irreplaceable one, and the bond between them deepens.

Let's see how this applies to Patrick and Anna, the couple we met at the beginning of this chapter, who came to couple therapy to renew their bond. They have been able to contain their negative cycle. When Anna complains, "I gave up my career for this relationship. Things are a little better now, but I still don't get the comfort I need," Patrick's initial reaction is still to withdraw. But then he looks into Anna's face and reaches for her, putting his hand over hers. "Yeah, well, giving comfort hasn't been my strong suit, has it?" he murmurs. They talk about how they know they need to learn to trust again and come close, but neither of them is quite sure how to do this.

So now they move into the second part of therapy—restructuring the bond by digging deeper into their feelings in Hold Me Tight conversations. "What happens to you right now when Anna gets upset and points out how much she has given to the relationship?" I ask Patrick.

"I don't want to shut her out," Patrick replies softly. "I know that doesn't work. But it's still very hard to stay here and listen. I hear the old song about how dissatisfied she is with me. I go into this, 'How can I ever be enough and make it up to her' thing. It's like we have found out in these sessions: I feel that I'm failing her, and I feel threatened. So my brain gets scrambled. Basically, I guess I get scared."

I want him to dig deeper into his feelings. "And when you begin to close down, like a little while ago—what's going on inside?"

"Oh..., hmm...I guess that is like a kind of despair—helplessness, maybe. I search for a solution in my head [he taps his head with his finger], come up empty, and then I cut off. Nothing to do here. Yes, it's like despair. If I am such a raw deal for her [he pauses for a long time], then this relationship is all washed up. [He looks up at me and Anna, and smiles an ironic smile.] No wonder I turn off, huh?"

"I'll say," I respond. "Anna gets upset, and you go into fear with no apparent solution. You hear that you can never make it and be seen as valued and precious to your wife. Despair and helplessness and a sense that the relationship is doomed; this is the kind of black wave that has you turning off and has Anna feeling so deserted. What is happening to you right now as we talk about this? Can you help her really tune in to this?"

What I am really urging Patrick to do is show Anna that he can fully engage with her. And he doesn't disappoint.

Patrick turns to his wife and looks her in the eyes. "Well... Basically I get crazy in these situations. I feel helpless. I hear that I cannot make you happy no matter how hard I try. And I have tried. I have. This helpless feeling is always sitting here [he touches his stomach] these days, just waiting to hear how disappointed you are in us, in me. I can run a huge company, but I can't hold on to you." He looks down, then looks up and leans forward. "I don't want to walk away from you," he declares. "You have probably needed my comfort many times. And I like that—that you need my comfort. I just get waylaid by my fears. All I hear you saying is that you got a raw deal in me!" Anna is looking at him, her face open and soft.

Patrick continues, "You are the center of my life, you and the kids. I don't want to keep getting caught in this dead end. I want so much for you to feel happy with me. But I need you to cut me a little slack here. Give me a chance to learn how to do this. Not assume that I just don't care if I don't always pick up on what you need. And I want you to see that I am trying. That there are lots of things I do for you. I guess I need some recognition. Need to know that you are not going anywhere, that we are going to stay together and work this out. To know that I am not such a bad husband after all... It is hard to feel this and tell you this. Maybe you don't care."

Patrick has become ultra-accessible here, actively helping his wife tune in to him. He is reaching out and asking for reassurance from Anna that his disclosures mean something to her. He is stating his needs clearly, letting Anna see inside him, and giving her a chance to respond.

But can she be there for him? I find myself thinking that to know how to reach for a precious one is the most basic skill in the dance of human connection. But we also have to know how to respond to another's need, to reach back in return, so that there is mutual engagement.

"I do care," Anna says slowly. "This is just so different. It's a relief to know that you are not just irritated or indifferent. Not that I want you to be scared like this. I didn't know I was that important to you! I just see you offering rational compromises that don't take my feelings into account. But you are scared!"

"Yes, I am," he responds.

I ask her how she feels, and she says, "I feel more connected to him. I hear that he wants to come closer. This feels better."

They have made great progress, but they still have some distance to go. He is out on the floor, but Anna is hesitant to let go of her mistrust and begin a new dance. If their bond is to grow, she also has to risk and reach, sharing her fears and needs in a different way. My job is to be a guide in this new dance.

"So, Anna," I say in our next session. "Are you feeling a bit more hopeful about your relationship?"

"Yes, I am," she replies. "But now I find myself still hanging on to those old glasses, if you know what I mean. He came and found me after a little tiff we had at home a few days ago and I knew he was taking a risk and reaching out for me, but I found myself holding back somehow and repeating my old mantra of how cold and uncaring he is. I guess it's still easier to just be mad than to open up to him."

"You got to the place in this relationship where you hunkered down and opened up the gun ports at the first sign of danger. So I guess that is hard to give up?" I ask. She nods. "How were you feeling after the tiff?"

Anna grimaces. "The usual. Totally alone. Like that 'here we are again' feeling. Ready to reach for my gun or turn into an ice cube!"

She looks at Patrick and gives a "What can you do" shrug. "But I am tired of being mad."

"When Patrick reached for you, even though you were feeling tired and alone, it was hard to respond?"

"Yes. It's hard to trust. To really believe that he is reaching and won't just disappear on me."

"Can you tell him?"

Anna turns to Patrick and says, "It's hard to believe that it's safe. I have felt so exposed with you. I get jitters in my stomach when I start to believe that you are here for me. I want that, but . . . it feels unsure. How do I know . . ." She is very still and silent for a few minutes, then resumes speaking. "How do I know that I won't get hurt again, left again? It's almost like I am scared to trust you now."

Patrick nods and leans forward. "Yes. I can understand that. We missed each other so many times, and both of us hurt. And sometimes I can be preoccupied or not very clued in, but I am trying to be here for you. I would like you to try to believe that. It's easier when you are not angry all the time."

Anna laughs. "Well, now I am angry just some of the time . . . when that voice in my head tells me to be careful. I think I am scared to hope, to really let that longing for you come out. If I do that and you are not there . . ."

"That would be unbearable, yes?" I ask. "Like falling into space? Devastating?"

"Yeah," Anna says, and turns to Patrick. "So I am afraid here. No—I am truly terrified. But I do long for you and I do need you to comfort and reassure me, give me some time to trust and to feel safe. I admire you for what you have done in these sessions. I want us to be close. Maybe I just need a little help."

"You got it," Patrick replies, beaming. "I will do my best. I am here." Anna smiles and reaches for his hand. "Well, right now, you are pretty much perfect! I guess I have to learn to trust a little."

He stands up and opens his arms and she moves into them. Anna has shared her fears rather than wrapping them in a bundle of rage, and she has found the courage to ask for what she needs.

After this session, I sit in my office and savor what has just happened. The words *reattunement, repair,* and *reattachment* go through my head. I feel happy. Connection cues joy, even when we are watching it happen in someone else. Our mammalian brain recognizes this as good, just the way we recognize the touch of the sun on our face. As a scientist and a researcher, I look at what just happened in my office and predict that by the end of our sessions Anna and Patrick will have a felt sense of connection, a safe-haven place, that offers the ultimate solution to emotional isolation and all the sorrows it brings.

At this point Anna and Patrick can do what securely attached dyads at age 3, 13, 36, and 66 can do. They can jointly create emotional synchrony. They are attuned to their own and each other's emotions and can empathetically respond to the softer emotions and attachment calls of the other. They are creating a loving bond right before my eyes.

This kind of event is powerful enough to undo years of mistrust and painful isolation, perhaps because of the flood of positive emotions it unleashes or perhaps because of the primal survival significance of the interaction. Whatever the reason, once these events begin to happen, couples can not only take a new and more positive path, they can reshape their inevitable disconnects into deeper trust that allows them to fall in love over and over again.

These kinds of events seem to render future miscues and disconnections unpleasant rather than catastrophic; separation distress, when it occurs, is manageable and resolvable. Partners can then help each other constantly broaden their response repertoire rather than scare each other into rigid, defensive postures. Can these kinds of positive experiences reprogram the brain and create trust and empathy, even when they have never before existed in a couple's life together or in either individual's past? I think so.

⌒

Once couples know how to open up, send clear messages, and respond to each other on an attachment level, then they have a secure base that helps them do this in other areas of a love relationship where attachment fears and needs get triggered, such as sex and traumatic injuries.

SEXUAL HEALING

At one point in therapy Anna and Patrick begin to address their sexual relationship. In this arena, roles have reversed from what they were at the beginning of their marriage; Patrick has become the pursuer who asks for more sex, and Anna is the one who retreats. Anna is now able to tell him, "I know I am kind of guarded with sex. And we have made some great changes, like more time for foreplay and more time after sex to hug, but I do still hold back. It's kind of strange, but I think for me it's a little like the feelings you have in our general relationship. You've said that you feel inadequate, and I guess I feel that way in sex. When you tell me your fantasies or say raunchy things when we make love, I feel sort of dismayed. I freeze up. I don't know how to be this sexy, over the top, hot woman that you seem to want. I don't want to yell my head off during orgasms — I am more of a quiet simmerer, I think. So lots of times I get distracted by a sense that I am just not hot enough for you." She sighs, looks down, and her voice goes very soft. "And maybe I never will be — it's just not me — so [she opens her hands in a gesture of helplessness] I want to avoid that feeling, I guess. So this has me holding back in lovemaking. But then you feel rejected."

This disclosure is a very long way from the negative comments she used to make to Patrick about their sex life. Those mostly focused on the suggestion that he was simply an adolescent stuck in constant horniness. Needless to say, the comments did not invite Patrick to engage in an open, exploratory conversation with her. But now, after her vulnerable remarks, she goes on to ask Patrick for reassurance that he sees her as a satisfying sexual partner, and he finds this reassurance easy to give. He also tells her that he was making raunchy comments because he thought this was part of being "hot" and making sure your lover felt desired. He admits that, on his side, any hint of a lack of desire on her part pushes him into a quagmire of doubt about whether he is loved.

As these partners become more securely emotionally connected and extend their newfound sense of safety into conversations about sex, their lovemaking becomes less a test of desirability and more a happy and satisfying affirmation of their relationship. This kind of Hold Me Tight conversation provides a platform of emotional safety for explorations of their unique sexual dance.

HEALING TRAUMATIC INJURIES

Hold Me Tight conversations also can be of enormous help in healing traumatic attachment injuries that result from single, shattering events (such as those discussed in the previous chapter). Hold Me Tight conversations, when focused on such injuries, promote forgiveness and the renewal of trust. A study I conducted with colleagues at my Ottawa Couple and Family Institute and at the University of Ottawa found that all distressed couples who came to us with a single attachment injury, such as an affair, could be helped in only twelve or thirteen EFT sessions with our experienced therapists. EFT increased their level of trust to the point of true reconciliation. These successful couples engaged in Hold Me Tight conversations that were centered around the injury itself and then, more generally, around their relationship needs. Partners became more open, responsive, and able to reach for each other. What's more, three years later they and their relationships were doing just fine.

The forgiveness version of this transforming conversation begins after partners understand how these injuries have affected their bond and how they can contain their negative patterns, such as demand-withdraw. They are then safe and engaged enough to go back into the hurtful incident. Here are the steps in such a conversation:

• *The hurt partner opens up and courageously communicates to his or her loved one the essence of their pain and loss. Each talks about themselves and their softer feelings rather than the flaws in the other's character. Core emotions and signals are clear.*

Alice tells Ben, "I have hammered you pretty hard, and I see now that this has played a part in keeping all this hurt going in our relationship. That night when our daughter got so sick, I felt so alone, so frantic, and so deserted. I just could not believe that you were not by my side. I never felt able or clear enough to really tell you this, and when I tried but couldn't get through, I just got exasperated. So I promised myself, 'Never again. Don't count on him to support you when you feel vulnerable. Don't.'"

• *The injuring partner works to listen and begin to tune in to the other's hurt, avoiding getting stuck in defense and denial and acknowledging that it is the wounded partner's hurt that matters more than the details of the event itself. The couple openly explores and shares what led to the injuring partner's inability to respond to the other's call for connection.*

Ben tells Alice, "You are right. I know I need to tell you that. I should have been there. I was so caught up in winning that contract. I was busy proving that I was the great leader I always wanted to be, that I was a success at last. I didn't understand how much you needed me. I minimized the whole thing. I just couldn't turn away from the 'success' thing, so I ended up letting you down. I didn't tune in. I was on the wrong channel." This kind of acknowledgment of the impact we have had on our partner really opens the door for deeper sharing and real healing.

• *Feeling heard and validated, the hurt partner can focus and articulate really clear messages about the injury, prompting the other partner to apologize with sincerity.*

Alice murmurs, "You say that you 'knew' our baby would be okay, but you weren't there looking at her. She looked so bad, so very bad. I believed she was going to die. It was like being hit by a truck. I couldn't breathe when that specialist told me what they were going to do. I had to give permission for them to operate, and you weren't there. It was like I didn't have a husband. It was just me, all alone, watching her die. And when I told you what it was like, you argued with me and told me it wasn't that bad. Then I was even more alone."

Now Ben's face mirrors his wife's pain. He is able to connect with his spouse's fear and hurt at his abandonment when she most needed him. He is engaged and shows her with the sadness on his face and in his voice that her pain hurts him. He expresses regret and remorse, and when he does it from this place of deep emotional engagement, it works. Ben whispers, "I let you down. I let us down. I am so sorry, sweetheart. I don't want you to ever have those feelings. To be so overwhelmed. I did not understand how afraid you were and how serious it all looked. No wonder you have been so very angry with me. I want to help you heal this. I will do anything to gain your trust again."

• *Once partners have shared their vulnerabilities, the stage is set for the penultimate step in an injury-focused Hold Me Tight conversation: the sharing of needs.*

Alice asks for what she needs to heal. "I still get scared," she says. "I watch our baby to make sure she is breathing right. I still dream about that night. In the dream, I call for you, and you don't come. I need to cry about this, and I need you to hold me. I need to know you will come this time." This time Ben responds. He holds his wife and tells her, "I will never let you down like this again. I want to reassure and comfort you. I will do whatever it takes for you to feel safe with me again. I will put us first." This comforting, open kind of connection acts as an antidote to the pain and fear of the injury and lays a new foundation for building trust.

• *In the final step, partners together create a new story of the injury. This story includes exactly how they discovered the way to heal their rift and, how to hold on to their new confidence in their relationship.*

In Ben and Alice's last therapy session, Ben tells me, "We have learned so much from this hard lesson. I never knew that closeness was something you made. I thought it just happened—or not. It feels good to know how much she needs me and that I can give her a haven that no one else can give. Now, *that* is what I call success!" They beam at each other.

In our lab, when we look at couples who successfully complete this process, what always stands out is, first, their willingness to explore deeper, softer emotions that lead to their discounted or unfulfilled attachment needs and, second, their willingness to risk turning back toward each other. As Ben told me, "It helps to know that there is a clear path through this kind of chasm; when you really understand the pain of it, it's easier to respond and help your partner heal."

The basic science of attachment gave us the secret to understanding these injuries and told us in a general sense what needed to happen to heal them. We could then build a model of the steps that can lead a couple from wounded despair to secure bonding. We now teach these steps to forgiveness, and the general Hold Me Tight conversation, as part of our relationship education program, Hold Me Tight: Conversations for Connection. The main insight here is that these wounds are abandonments that spark life-and-death survival scripts and attachment panic.

Our research into the key events that change distressed relationships into more secure bonds tells us that if we can understand the drama of attachment and how we deal with disconnection—and if we can learn to accept and call out our deepest attachment fears and needs, and if we can respond to these calls with attuned care—we can, with purpose and deliberation, grow our deepest bonds across a lifetime.

For many of us this is a startling revelation. We do not have to travel through this life alone, relying only on ourselves or the whims of mysterious love. These megawatt emotional conversations rebuild trust and lead us to new levels of intimate connection. To know, at last, how to grasp, shape, repair, and renew our most important adult relationship, the relationship that, if we understand it, can sustain and nurture us throughout our life—what can be more important than this?

EXPERIMENT 1

Sit quietly and imagine yourself in a conversation with a loved one whom you do not or did not always feel safe with. See if you can remember a time or specific incident when you felt disconnected and hurt in this relationship. Ask yourself what threat was present.

Was it the threat of imminent rejection—that is, learning that this person did not value you or the connection with you? Was it the threat of being abandoned or deserted—learning that this person could turn and walk away, leaving you bereft? Was it the threat of learning that this person judged you as unimportant or unacceptable? What, to you, was the most catastrophic thing this person said or did? See if you can pinpoint the exact moment that hurt the most.

Ask yourself what you needed at that point that would have turned the hurt and fear around. What did you long to hear or have that person do? Now, for a moment, imagine that person magically tuning in to you and doing just that. Give a name to what you would feel—for example, intense relief, deep comfort, dissipation of fear.

But suppose this magic didn't quite happen and you had to help the other person figure out how to respond. Imagine what it would be like to tell this person about the threat you felt in that situation. and the message or action you needed to receive in that moment. See this person's face, and see yourself sitting opposite him, and beginning to speak. If you can imagine yourself doing this and being able to send a clear message while keeping your emotional balance, that is great. You have just primed your attachment system and rehearsed your part of a Hold Me Tight conversation.

If you had a hard time with this last piece, see if you can determine what most got in the way of your giving a clear message about your needs and fears in the conversation. Here are some of the blocks that people identify:

- I cannot keep my focus. It is hard to stay with the feelings, so I change the subject or get abstract and tangential.
- My feelings reach flood level when I imagine sharing like this with this person, and the risk suddenly seems too great, so I shut down.
- I find myself flipping into anger and blaming—proving this person wrong instead of sharing soft feelings.
- I knew when I imagined this person's face that this is all pointless, and I wanted to give up and run or hide. This is too hard.
- I find myself telling this person that I will never trust him again. I will never let him hurt me again. I want to protect myself. I refuse to tell him my needs and fears until he proves worthy of my confidences.

If you have never seen or experienced a Hold Me Tight conversation, then this thought experiment is a beginning, a way to explore it as a possibility. If you do have a prototype for this conversation, then it is a chance for you to hone your sense of this exquisitely powerful interaction.

EXPERIMENT 2

Choose a small hurt inflicted by someone you depend on. Ask this person if you could just talk about the event. Say you want to see if this telling would be helpful to you. Say that it is fine if she cannot or does not wish to respond and that you are not trying to blame her or make her feel bad. If she agrees to listen, try to pinpoint your specific moment of hurt in very simple language while staying soft and open. Stay with your task and see if you can do it, whether she responds or not. Write down what this was like for you.

PART FOUR

The New Science Applied

CHAPTER 9

A Love Story

I remember thinking how often we look, but never see…we listen, but never hear…we exist, but never feel. We take our relationships for granted. A house is only a place. It has no life of its own. It needs human voices, activity, and laughter to come alive.

— *Erma Bombeck*

We need new love stories. This chapter shows one couple, André and Cleo, moving from breakup to bonding. We see them in therapy, at home, and in their car as they grapple with their fears and longings. We watch them struggle to find their way to renewed closeness and connection. The key to a loving bond lies right before their eyes, but they do not see it until they have been taught how to look. The same is true for all of us. This is not a mythic tale of romance but a down-to-earth story of love made possible by the love sense that is now available to us all.

243

André pushes his dark hair out of his eyes and stands looking out at the November rain. It is starting to get cold. After ten weeks of feeling the heat in the Middle East, where he was consulting with computer companies about setting up new programs, just the sound of the wind and the rain chills him. But it is the change in his relationship with his wife, Cleo, that really leaves him chilled. For six years, they had been partners. When he first met her at the local gym, she had been recovering from an injury to her back, and he had slipped into taking care of her and helping her get back to the high school teaching job she loved. He had become her friend and then her lover.

Who would not want to take care of such a beautiful woman? He was just happy that she was with him. What did he know about women? They were a mystery to him. He had never understood the whole romantic thing. His parents had been cold and aloof, sending him to his room if he became upset and teaching him that talking about his feelings put an unwarranted burden on others. He had not dated much, and when he met Cleo—small, blond, with kind eyes and a soft smile—he could not believe that she liked him.

And so he had been so very *careful*. They had never had a fight in all the years they had been together, even though Cleo is kind of volatile. She just says what she thinks. She doesn't hold back. But they had been best friends. Sometimes, over the years, Cleo had complained that there was a lack of passion between them, but he could still feel the total disbelief and panic that hit him when she had e-mailed him, just as he was about to fly home, that she had started an affair with a colleague at work.

The memory of arriving home, running into the house, exploding at Cleo, and pulling apart the bed where he believed she had made love with her colleague feels more than a little foreign to him. Who had done that? Somehow he had managed to stop her from walking out on him, and now, weeks later, she has even agreed to go for couple therapy. But it feels to him like they are clinging to the sides of an abyss. If they let go for an instant, they and their relationship will fall through space and shatter.

THE SPIRAL OF DISCONNECTION

André waits for Cleo to come back from the kitchen with the after-dinner coffee and turns to her as she places the cups on the small table in front of the couch.

> *Cleo:* [She smiles at him.] What are you thinking about, staring out into the rain?
> *André:* Nothing. Just watching the rain. [He notices that she is pursing her lips, and then she frowns.]

His brain reads the message in her face in 100 milliseconds, and in another 300 milliseconds he literally feels her irritation in his own body, courtesy of his mirror neurons.

> *Cleo:* [Her voice is clipped.] You were going to bring in the summer furniture. Now it will be starting to rot.

Cleo instantly reacts to "Nothing." She has reached out to André, asked him to share himself. She is disappointed and annoyed by his response but wants to stay safe and at a distance. So she focuses on the word rain *and expresses her annoyance at his aloofness by complaining about a chore he has neglected. The danger here is that they will get stuck in the content issue of chores and miss the attachment cues. They will assume the fight that is brewing is actually about rain and chairs or his forgetfulness. They won't realize that the fight is really about lack of connection and what kind of bond they are going to have.*

André: [His face goes tight and flat.] I can do it tomorrow. I never said I would do it immediately.

His amygdala picks up on her annoyance. He defends and freezes up. She is going to feel more shut out now as he stonewalls her. The main emotional message lies in how he speaks, not the words he says.

Cleo: No—you said you would do it yesterday. I guess the promises you make to me don't matter much.

She hears in her mind "I and my wishes don't matter much to you. As I invite you to share, you do not respond to me." All this has taken about 10 seconds. Attachment panic is now hijacking both their brains.

André: [He squares his shoulders and speaks in a very clipped tone.] You are not the one to talk about broken promises, Cleo.

Cleo: [She flinches and shifts down into logic.] I do not want to have a fight about the damn furniture, so you don't have to get all defensive.

The fight, as always in these suddenly stiflingly hot conversations, is not about the surface issue, the furniture. It's about the quality of the connection and the perennial question "Are you there for me?"

André: I am not defensive! Who said anything about a fight? You are just angry *again*. [He turns away from her and defends himself by lashing out at her with a critical comment.]

André's heart rate has skyrocketed. He is flooded, moving into separation distress. The special pathway for attachment panic has just lit up in his brain. His neurons scream "danger" in the same way they would if his car were approaching a cliff. This happens for Cleo, too, as her sense of helpless abandonment grows. She is probably going to deal with this by attacking him. Then she will feel some sense of control.

> *Cleo:* At least I don't suck everything up so that I almost disappear. I don't bottle stuff up at all.
>
> *André:* What are you talking about? You don't bottle anything up? You are the one who just suddenly went off and had a secret affair. I don't see the point in talking like this. We just get all tangled up in feelings here, and you get mean. [He walks back into the kitchen.]

This couple's dialogue—which consists of pursue, complain, and criticize, followed by defend, withdraw, and stonewall—is up and running; each pulls the other deeper into primal panic and loss of emotional balance. Both are lost and feeling vulnerable. No one is clear as to what the fight is about or how it got going. Danger signals are everywhere.

> *Cleo:* [She follows André into the kitchen and suddenly sees pain in his face. She softens her voice, changing the emotional music.] Well, you are so distant. You never share anything with me. I didn't mean to hurt your feelings. [She walks across the kitchen to stand near him.] There is no response from you. Maybe that is why there is so little passion in this relationship.

Courtesy of her mirror neurons, Cleo feels André's pain and moves from misattunement and disconnection into a moment of attunement. She names the problem in attachment terms and in terms of sexuality.

> *André:* [He closes his eyes, leans against the sink, sighs.] I

told you. I am not sure how to make that happen. We were always friends. I guess I am a bit of an island, and passion was never my thing. I am a chameleon. I don't even know how I feel most of the time. When you were happy, I was happy. My emotions were always a bit of a mystery for me. [Cleo reaches out and touches his arm; his face softens.]

André feels soothed and so explores and discloses his confusion and his sense of being cut off from himself and from others. He is able to reflect; his prefrontal cortex can come online now that his amygdala has stopped flashing red for alarm and commandeering all his energy and attention.

Cleo: To me, it does feel just like you said—you are an island. [Her tone changes now, taking on an edge.] And that really upsets me because it seems like our relationship is no different from the ones you have with your buddies. I am just another buddy for you. When I told you about the affair, I thought you would just leave.

André: [He speaks in a hard tone and with a still face.] Come to think of it, I don't know why I didn't leave. I am not sure what is holding us together. You are angry all the time, and I . . . Well, we both agree that I am an island, so . . . [He throws up his hands and looks at her with a still, closed face.]

Suddenly André, who was opening up a little, hears blame and moves into defensiveness and helplessness. His fear of rejection is primed. He shuts down, and his still face will make it almost impossible for Cleo to keep her balance. Their brief moment of safety is lost. Now she will protest his retreat. If he were to tune in to his deep emotions right now he would find a sense of failure and loss waiting for him. It seems easier to give up.

Cleo: [She tears up and speaks in a loud tone.] And what is that supposed to mean? We never work out our problems because you won't hang in with me. You bail. Just like you are doing right now. If you don't know why you are here, then maybe you should get out. Leave, why don't you? I am sick of having to chase you all the time just to get some kind of reaction.

His apparent ability to give up on their relationship hits her raw spot, her fear of abandonment, and she reacts in rage. She hears all her worst fears confirmed. She cries out from her own pain and to try to get him to see her.

André: [He is quiet and apparently calm.] Well, you sure found a way to get a reaction, didn't you? Screwing someone else will do it every time. The first real test in our marriage, the first time we are apart, and you screw someone else for a week straight and announce you are in love with him and you are leaving. It was like I was held at gunpoint. Fight or lose everything.

We know that André's calmness hides a physiological storm, and he naturally goes back to an unresolved moment of hurt, the moment when he thought he had lost her. The life-and-death imagery of "gunpoint" is typical. This kind of attachment injury, unless and until it is healed, will be continually triggered and spark this couple's negative cycle again and again. This kind of injury is indelible. It will not fade with time. The only way out is through.

Cleo: [She turns away and flops down on the kitchen stool.] I didn't mean to do that. It wasn't planned. It just happened. You didn't need me anyway. And now you are never going to forgive me, so we are doomed. [She turns back toward him and puts an edge in her voice.] And it's not fair, because you were part of what happened.

André: [He turns his back to her and uses a logical, calm tone.] We were talking about having a baby. How exactly was I supposed to know my wife was about to leave me for someone else? It was like you just wanted to hurt me. And now it's all about you... Some part of me says I am stupid to be here. I should just put my jacket on...

Cleo: [In an angry voice, but with hurt all over her face] So now I get punished forever. Is that it? We had problems. Problems with intimacy. Problems with sex. It's like we were both always waiting for the other person to make a move, to show some desire. We got that book on premature ejaculation, and then we didn't read it. Then I got depressed, and you didn't want to talk about that. The affair was a way out.

Too many hot messages, coming too fast. All the signals are ambiguous and distorted. She shows hurt but expresses anger. He stays with the danger cue of her anger. Her comment that the affair was a way of dealing with despair and loneliness is classic and accurate.

André: I made moves. Maybe they were too subtle for you. But you'd miss them, and then I'd give up. And now your "way out" has totally smashed the foundation of our relationship, the trust between us. I don't know what I can do about that. There is nothing I can do. You told me very clearly that you were not even attracted to me anymore. I can't live with that. I can't make you happy. Let's just leave it at that.

André is probably right. His tentative sexual moves were not obvious enough to invite Cleo into her desire. We know now that female desire is often responsive in nature. If you believe there is "nothing you can do" to fulfill your lover, the sense of failure is crushing. He is dealing with this sense of failure the only way he knows how, by turning away and by trying to express his helplessness, but she cannot take this in and hear it over his threats that he is leaving. Full speed ahead, emotions on full throttle. The way each deals with feeling vulnerable pushes the other over the cliff.

> *Cleo:* [In an accusing tone] You don't try. I never see you desiring me. You let the fire go out, and now you blame me for turning somewhere else for warmth. [She rushes from the room, weeping.]

What just happened here? If I asked Andre and Cleo, they would shake their heads and say they do not know. The steps in this dance confirm each partner's worst attachment fears and that it is dangerous to move into soft emotions with the other. Each reiteration shrinks the safe ground where trust and empathy can grow. The other person begins to look like an enemy. But if we can grasp the nature of the dance, the steps we take that put our partner off balance, and if we know how to build a safe haven with our lover, this can all change.

Four months later, André and Cleo have found an EFT therapist and have begun to understand their relationship as an attachment bond. They are able to come together to control their negative dance and grasp that their personal ways of dealing with emotional insecurity trigger the other's fears and trap both of them in lonely desperation. They are learning to stand together and to help each other with their vulnerable feelings and raw spots. Now they can move into a Hold Me Tight conversation.

A CONVERSATION FOR CONNECTION

André walks into the living room, bringing the after-dinner latte that he has just made for his wife. He puts the cup down in front of her, sits down beside her, and looks directly into her face.

> *André:* [Softly] How are we doing, then? Us—as a couple, I mean.
>
> *Cleo:* [Smiling] Better, I think. We don't get as stuck in that spiral where I poke to get a response from you and you tune me out or go at me. We don't seem to freak each other out as much. Seems like we feel safer together. [She pauses.] What is funny is that I feel more attracted to you than before, and we aren't doing anything differ-ent—technically, I mean—as lovers.

This is different. She has an overview of the dance that mired the relation-ship in distance and defense and left them both aching for connection. He seems more open now; he approaches her, asking about her feelings concern-ing the relationship.

> *André:* Well, it helps that we can talk about stuff. I was so anxious, kind of shy. I was always so worried about the performance part, living up to your expectations. Being sexy was like trying to play and respond to someone when you are in a minefield. And then you would get disappointed and not want to come to bed or even flirt, so... We were never in the same place at the same time. Foreplay sucks if you're scared. [They both laugh.] But it feels like you are starting to trust me more.

This is better. We know that just being able to name our deep emotions seems to calm us down, and then we're able to attend to the recurring scenarios that unfold.

> *Cleo:* [Picking up her coffee cup, speaking softly] Yeah. Trust was never really my thing. It's too easy to get hurt. So I protect myself. I think my family taught me that. But you are pretty guarded, too, you know. You don't let people in. You are the distance expert in our spiral.
>
> *André:* [He moves to lean away from her and speaks fast.] We are not going to have a discussion about my flaws now, are we? You can be pretty critical, too, you know. And in fact, I am not being distant right now...

Oops. He is triggered here. Emotion is fast, but Cleo keeps her emotional balance. She doesn't get swept off her feet and caught up in the destructive spin. She pulls him out and helps him get his balance back.

> *Cleo:* [She reaches out and puts her hand on his arm.] Yes, yes, okay. André, you are right. You are not being distant now. And I didn't mean to hurt your feelings. I am the poking expert in this relationship; being critical is easier than showing my soft feelings, that is all. I don't want to be critical. I appreciate your bringing the coffee and asking me how I am feeling about us. I appreciate that. There was a time when you would never do that.
>
> *André:* [He leans back in the chair.] Damn right I wouldn't. If you are tiptoeing around wondering why your wife is with you at all, when you don't have any idea what you are doing as a husband or a lover, and you think it's just a matter of time before she bails on you, then the last thing you do is ask a question like that. [His voice

slows and drops.] I guess I was always waiting for you to leave—and then you did!

André can now put his experience together into a coherent story; one where his anxiety and catastrophic expectations got in the way of being present with his wife.

Cleo: [She reaches for both his hands and holds them.] But you fought for me, and we are here now, together. Yes? [She searches his face and holds his hands tighter.] And you asked.

André: [He smiles at her.] Yes, well, somehow, in the last few weeks, we seem to have gotten a lot safer together. Or I am less of a chicken or something. [They both laugh.] It helps to know that we got caught in the same dances that everyone gets caught in and that I am just an average schmuck figuring out how to be close. Just like everyone else. Not the only one who doesn't get it! And it feels good to be able to get hold of my feelings, so I can tell you what is happening, rather than always waiting for the Great Inevitable Rejection and for this cloud of aloneness to swallow me. It's kind of liberating, in fact.

Wow! André is talking like a securely attached partner here. He has a sense of confidence, of being able to work with and manage his emotions and his relationship with Cleo.

Cleo: My sense is that it is still hard for you to tell me that softer stuff. That is kind of sad. I like it when you take risks with me and let me in like that. That is what I have always wanted.

André: [He blushes and looks down.] It's pretty heavy, that feeling, that fear. It's almost like a panic attack for me. My instinct is to hide. It's pretty hard to speak from that place. [She nods and leans closer to him.] It's like this great torrent of feeling about not being good enough for you. I don't want to feel it. And the idea that you might want to hear that from me, that is really new, Cleo. I never had anyone to share that kind of stuff with. Never had that. I can hardly believe it, really. [He looks up into her face.]

Cleo: [Very quietly] I get that. But when you take risks like this, then I know I matter to you—that you need me close. Otherwise I just feel shut out. If we can comfort each other, then think of the possibilities!

André: [He passes his hand across his forehead and sighs.] I really am trying, Cleo. It still feels like dangerous territory. This voice in my head tells me that one day you will decide to trade me in for a better model, so I should hide any uncertainties. [Cleo opens her mouth to speak, but he stops her.] I know, I know. You are going to tell me that you turned away because you were lonely—because you couldn't "get" me—not because you didn't want me. [She nods at him slowly, and he breaks into a smile.] I still see myself as a pretty replaceable guy.

Cleo: [Speaking in a very deliberate tone] I love you. And we nearly lost each other. For me you are special. My André. Unique. You don't have to be some perfect person. The more you turn toward me, the more you will see that I will try to be here. We are going to kick that guy with the voice—the not-good-enough guy—to the curb. But I have to be able to see you, André. It's hard to love a stranger—not safe. Do you get that?

André: I get it. I get it. And this takes the pressure off me. It helps my confidence when you say this kind of stuff.

André and Cleo are moving into a Hold Me Tight conversation here. He is open and sharing his attachment fears. And she helps him with those feelings. This is the beginning of a more secure emotional bond.

Cleo: It's like last week, when you partied too much and couldn't drive us home. I didn't flip into anger because you opened up and told me you felt bad about yourself for doing this and not getting me home on time. I didn't need to yell to show you how much you hurt me, because I saw that you already cared about it. So I could feel some empathy then. I guess my angry messages have been part of your shutting down.

André: Yeah. You even ran into a pharmacy and got some ginger stuff for my stomach. I couldn't believe that. It's like we are more of a team and I can screw up sometimes without the sky falling in.

Cleo: [She looks down and begins to twist her fingers together.] I know I need to ask, too. To open up. Share more. In a way, your fears about me were right on. Even when we got married, I told myself that I always needed a plan B. My mom used to tell me, "You are the only person you can rely on." With my mom, I think I learned that if you show where you are vulnerable, all that happens is that people know how to hurt you and you will be burned again. So the only safe way is to just count on yourself. [André nods and puts his hand on her arm.]

Cleo can put the key messages she got from her family about depending on and bonding with others together with her experiences of closeness and see how these shaped her main strategy with romantic partners. As she does this, she opens the door for change. Research says we can change these strategies; we can learn that we have more options.

> *André:* I know I felt like it was dangerous to depend—better just to count on myself—before I met you. I had to do this; there was a void. No one was there. I told myself that it meant I was independent, so it was okay. So what is that like for you? Is it hard for you to feel that way, just counting on yourself?

Neat. Now that he feels safe, he can be curious and explore her reality with her.

> *Cleo:* [Very softly and slowly] It's draining. Too hard. Really lonely. I don't want to be married and have a plan B in my pocket. It was kind of a habit, thinking like that. And anyway, it doesn't work, because at the same time, I would get enraged and throw myself around to get your attention, to make you open up to me. I don't want to spend my life being mad or feeling lonely, like my feelings don't matter to the person I am with. I need you close. It's too painful to feel like you are all by yourself when you are in a relationship. I want to let down my guard, but it's hard to reach for you when I am not sure that you will respond to me. [He strokes her hand. She smiles softly.] It's only taken me six years to begin to figure this out!

Cleo reciprocates André's sharing of fears and soft emotions. Her messages are clear; her nonverbals match her verbal message. And she goes one step further and asks for what she needs. Cleo's open sharing stirs André's empathy and makes it easy for him to tune in to her.

André: That's okay, sweetheart. We are figuring it out to-gether. I don't want you to be lonely, to feel like I don't care. Seems like we need to make these signs that we need each other really clear and easy to see, like we are now. I was so caught up in my own anxiety, I think I just used to miss most of your signals that you needed close-ness. I don't want to miss your cues, Cleo. I like it when you tell me what you need. I can be there. [She reaches over and hugs him for a long moment.]

This is the touchdown. He shows her he is there for her and offers her a safe haven and a secure base. This kind of response changes bonds and the people who make them.

Cleo: [She leans back on the couch, reflecting for a mo-ment.] So if we can work this out, then maybe no one has to hide or poke [she pokes his arm] or be lonely, right? [She wipes away a tear.]

André: Hmm...[Long silence. They both drink their cof-fee.] But one thing sort of confuses me. If I open up more, how will that work for us in bed? I thought men had to be mysterious to be a turn-on. The thrill of the unknown. Women love mystery. If I can't make the sex work between us...[His face is suddenly a picture of doubt and helplessness.]

André is right. Sex helps keep a bond alive; if there is no synchrony in this physical connection, it's harder to stay emotionally connected.

Cleo: I am not so sure about that mystery thing. In movies, maybe. In real life there is too much at stake. And I was too angry to make love. After a while, I couldn't respond to your moves. I didn't trust them. When I feel close to you, sex works for me. I want to feel desired. That is the biggest turn-on for me. But my sense is that this panic-and-shutdown thing is a big part of the sex issue for us. You would stay careful. For one thing, you were worried about losing your erection all the time. Well, if we can just relax a little and play, I can help you with that, too. [She giggles and opens her eyes wide at him. She whispers in a sexy foreign accent.] "I can help you, mister." [Giggles again.]

As Cleo and André relax and shape a safe connection, we know that they can communicate their sexual needs and desires more effectively and also let go and simply indulge in erotic play.

André: [He laughs.] You can laugh, but it's hard for me to play in bed. Playing is unpredictable. It could sweep someone away. All this worry and caution just comes and envelops me. But it's getting better...

Cleo: Is there something I am doing that is helping?

André: Yes. When we take our time, it helps me. When you touch me lots. That kind of soothes me, and I start to think, "It's Cleo, and she does want you," and then the carefulness kind of shrinks. [André laughs.] The carefulness, I mean—not anything else! [They both laugh.]

Cleo: When things got bad between us, I think I stopped all the hugging and touching...

André: Right. And then that voice in my head just got louder—the "she doesn't love you, you idiot" voice. And sex got more pressured, more difficult. It was like a test every time for me. So I would avoid it. [He goes quiet for a long moment.] I think I need your touch. It grounds me somehow. I instantly feel like I am not alone. It was like I had missed that my whole life and didn't know what was missing. [He tears up.] Then I found you, but I kept waiting for you to disappear. I need your soft touch.

Here André completes his part of the Hold Me Tight conversation, telling his wife what he needs from her to stay open and present both in and out of bed.

Cleo: [She leans across and kisses him.] I will be your ground, love. It takes so much strength to do what you are doing right now. You are the man for me. [After a minute she speaks in a mischievous voice.] So now will you play?

He laughs, gives a joyful roar, and reaches for her.

A CONVERSATION FOR FORGIVING

A few weeks later, Cleo and André feel safe enough to talk more openly about the wound that still festers in their relationship: Cleo's affair. This Hold Me Tight conversation focuses on the hurt from a key event and how it continues to block trust between the couple.

André and Cleo are driving to a friend's country cottage for the weekend. They had tried to do the Forgiving Injuries conversation from the book Hold Me Tight *the night before but had been interrupted by a call from Cleo's mom. They try to begin again. André is driving.*

Cleo: I thought it was good that we started to talk about it last night. Even if we didn't get that far. It's hard for me to hear how devastated you felt. That was the word you used, wasn't it? [She looks at him, and he nods deliberately.]

André: That was the word. [He keeps his eyes on the road.] It was my worst nightmare come true. Sex had become a minefield between us. You made it clear that I was failing, disappointing you in bed, and so...you had sex with someone else when I was away, out of the picture. Like our marriage meant nothing to you. I had lost you. The only thing that stopped me going crazy on the plane home was that I got enraged. Destroying the bed felt good. [After a long silence, he continues.] I thought it was just guys that were supposed to be all hung up on sex being great. [Another silence, and he moves the car into the slow lane.] When I think about this, I start to wonder if I am a fool for still being here. Even though we seem to be better together, this still kills me. In fact, if we are going to talk about this, I think I had better pull over. [He pulls the car into a little lane looking out onto rolling farmland. He stares out the windshield and speaks very quietly.] I can't ever go through that again, Cleo.

André now talks about this in a way that acknowledges and exposes his vulnerability and pain. He does not assign blame to Cleo. This ability to express your pain in attachment terms is the first step in the EFT model of forgiveness.

Cleo: It wasn't about sex.

André: [He turns to face her.] Excuse me? Seems like that was a pretty big part of it to me. It was part of your plan B, I think.

Cleo: [She turns to face him.] You are angry whenever we

talk about this. I get that; you have a right. But you know what was happening with us. We were totally cut off from each other. We were either distant or fighting. It wasn't part of a plan, unless you call it a plan to want a way out of hurting and being so alone. I wanted to feel desired, and I just fell into it. When I e-mailed you, I expected you to just agree, to say you didn't care. I told myself that that would be for the best. I was so angry at you as well. I was lost in all these feelings. But when you came home, everything changed. Being desired wasn't the big solution. It was you I wanted that from. It was an escape, an illusion that just fell apart. I know I hurt you very badly. Maybe some part of me wanted to create a storm just so things would get clear, just so we could break out of our ways and move somewhere.

Well done, Cleo. She doesn't get defensive. She acknowledges André's pain and explores her own actions. She sets out how she moved into the affair. When she does this, she helps him make sense of her actions. She begins to be predictable again. Step two in the tested EFT model of forgiveness is unfolding.

André: [Very quietly] I am angry but mostly I am hurt, hurt, hurt that you would do this. Yes, we were cut off from each other, but still...I would never have done that to you, Cleo. Maybe I am not such a great lover, but...

André is protesting her disregard for him and their relationship. We hear the three elements that seem to constitute what we call hurt feeling: reactive anger, loss or sadness, and fear that she could so dismiss their relationship and reject and abandon him. But he expresses this hurt softly and allows her to stay connected and respond in a way that can help him heal.

Cleo: [Speaking very quietly, in tears] I was very desperate, André. I could not go on the way we were. I got all hung up on the sex aspect, but that was just a symbol for the whole "alone" thing. Now that we are more connected, we are working out the sex stuff just fine. I guess I just wanted to feel that someone loved me, wanted me. It is kind of pathetic, but there you are. He flirted with me, and I just let it happen. But I think I was also so angry with you for shutting me out so much. When I wrote the e-mail I was mad. I wanted to show you that I could...I don't know...that I could get to you even behind your wall. The minute you came home I knew I had made a terrible mistake. I just want us to find a way through this, find a way together.

André: I can't respond to your anger, Cleo. I just get totally overwhelmed by it. So yes, I have to shut down. I guess I did shut you out. [There is a long silence.] It does help for you to tell me what you were thinking. It makes it more tangible, more logical somehow. I guess we were both lost and desperate. You kind of acted out my worst fantasy. Maybe something dramatic had to happen, I don't know. All I know is that I can never, never go through that again. I felt like our relationship was nothing to you—nothing. An e-mail message! If I hadn't found my rage, I would have come apart. No—I came apart anyway. [He cries.] How could you do that? And if I am not the greatest lover ever, will you do that again? Some part of me wants to just run when I touch this hurt. I have never felt so small.

André does what all the partners who succeeded in healing their relationship in the EFT study on forgiveness of injuries did. He risks opening up his softer feelings and showing her his pain. He tells her the impact her actions had on him and how hard it is to risk depending on her again. He can do this in part because of the help he's had in finding and making sense of his emotions and in part because he loves his wife and is willing to risk reaching for her again.

> *Cleo:* [She turns to face him. She is weeping and reaches to put her hands on his arms.] I am so, so sorry, André. It was the tackiest thing I have ever done. You didn't deserve that. I just wanted a way out of feeling so lonely. [He looks up into her face.] I do care about your feelings. I don't want you to hurt. I feel sick when I see how much I have hurt you. Ashamed and sick. I will do whatever I need to do so you can feel better and begin to trust me again. I am sorry, my love. [He grabs her, and they hold each other and cry for a while.]

This is the gold-star apology. Now he can see clearly that his hurt causes her pain. She expresses sadness and shame in a deeply emotional way, a way that moves him. This emotional connection builds new safety where before there was only danger and pain.

> *André:* [He leans back against the car door.] I guess I need you to tell me that we will keep on working on our relationship and our physical connection together and that you will be with me in that. I need reassurance, I guess. And when this feeling comes up about all this, I need to be able to come to you for comfort, to hear you say the kinds of things you just said. It helps to know that you have moved to another office at work so you no longer see him. But it's not really him that haunts me. It's that you didn't think of me—of us—and you sent

that e-mail. I felt so dismissed, so unimportant to the most important person to me. It was terrifying. I need to know I matter to you.

André's emotions and the needs behind them are now crystal clear, so he can send a compelling, coherent message to his lover about what he needs to move into a sense of felt security with her.

> *Cleo:* [Very softly] I made a terrible mistake. I will do anything to help you with this hurt. You are the person I want to be with. I will not turn away from you again. I want us to be together. I do not want you to be afraid with me. I am here.

This is the essence of a Hold Me Tight conversation. If this couple were in our research projects we would rate each statement in this conversation, noting the depth of the emotion expressed and whether André can integrate it so it makes sense, whether he can stay open and reach for her rather than lapsing into defensiveness or anger, and whether she can come to meet him with compassion and caring. Watching this couple's conversation, we would predict that they would, at the end of therapy, reach new levels of happiness and trust, levels that would persist in the years to come. Physiologically, they are in sync; emotionally, they are open and responsive to each other's cues.

> *André:* [He smiles very slowly.] Well. Well. I guess that is good, then! Somehow this feels much better. I don't know what we did, but I feel like a weight is off me. Maybe we really can hold on to these good things we are doing; maybe we can. If we can heal this one...

He is right. As they undo the old hurt and replace the pain with a sense of closeness, they renew their bond. And they become increasingly confident that they can shape their relationship and steer it through any crisis. They have a safe haven and a secure base together.

> *Cleo:* [She dries her face with her hand and smoothes out her coat.] We are healing it. We are reaching and learning. It feels like a new place altogether. If we can come out of this crisis and learn to hold on to each other, well, we can do anything.
>
> *André:* [He smiles.] Okay, then...Let's get up to the cottage and start our weekend. You and me. [She smiles back at him. He starts the engine.]

EXPERIMENT

What was it like for you to read this love story? Did it seem strange or foreign? Did it remind you of moments in your own relationships?

Were there parts of it that you think might have been pretty hard for you to do in your relationships?

What would you like to have said to André and Cleo as they let go of the support they received in our sessions and went off to build their new relationship together?

CHAPTER 10

Love in the 21st Century

here is the deepest secret nobody knows
(here is the root of the root and the bud of the bud
and the sky of the sky of a tree called life;which grows
higher than soul can hope or mind can hide)
and this is the wonder that's keeping the stars apart

i carry your heart(i carry it in my heart)

—*e. e. cummings*

All of us as parents hope to keep that sweet sense of connection we feel when our child struggles to toddle toward us, eyes wide with the delight of reaching our outstretched hands. So I am ridiculously touched when my daughter, now a beautiful, cool twenty-one-year-old, actually asks me out for coffee. She even picks the small, intimate coffee shop just down the road, where we can, as she puts it, "have a real convo." I feel smug knowing that in the lingo of the much-younger-than-I-ever-was folks, that last word is short for "conversation." All right. The coffee is perfect, the little café cheerful, and the frothy lemon cake delicious.

Once the small talk is over, I focus my eyes on the patterns in the foam of my latte, and I confide in her about a small victory in my work life about which I feel ridiculously proud. That's when the strange sense that I am speaking into a vacuum hits me. I look up, and my daughter's thumbs are flashing across the small screen in her hand. She is texting one of her friends, and I am talking to myself! I straighten my spine, mentally put on my red-and-blue superhero suit, and in the interest of human connection on planet Earth, I roar, "It's me or the tiny screen, sweetie. Choose. I won't talk to you when you're not here."

In these pages, I've shown how the quality of our relationships with others is the bedrock on which we build our existence. Our closest love relationships shape who we are and, more than perhaps any other single factor, shape our life story. Happiness experts, such as psychologist Ed Diener of the University of Illinois, tell us that our relationships are the strongest single predictor of human joy and well-being. Ever since social scientists started systematically studying happiness, it has been resoundingly clear that deep and stable relationships make for happy and stable individuals. Positive relationships also make us more resilient, advance our personal growth, and improve our physical health.

But relationships matter on a grander scale as well. The ways we tune in to and engage with others sculpt the very society we live in. Secure connection with loved ones helps us be open, responsive, and flexible, and that, in turn, makes us inclined to perceive the world as kinder, safer, and more malleable. It gives us the capacity to look outward, to see the broader universe, and take a more active role in it. Positive relationships make us more apt to be community builders—creative workers, good leaders, and caring citizens. A civil society depends on connection with and trust in others—on what primatologist Frans de Waal calls the "invisible hand" that reaches out to others.

Yet increasingly, we are not here for each other. Ironically, just as we finally crack the code of love, we seem to be doggedly building a world where such bonds are less valued and harder to make and maintain. As individuals, we are ever more cut off from each other in a fast-paced and socially fragmented world. Historian Ronald Wright has a harsher assessment. Modern civilization, he argues in *A Short History of Progress,* is a "suicide machine." Indeed, the "flight recorders in the wreckage of crashed civilizations" indicate that Western industrialized society, with its promulgation of narcissisim and greed, is in free fall.

The only hope for *Homo sapiens* is to "know itself for what it is." With such knowledge we can build a society that fits with and complements our most human, most humane, nature. As Aristotle said, "What a society honors will be cultivated." It is time for us to understand, honor, and cultivate the deepest relational elements in our nature. We must build on the social capital that is at the heart of any civilization that merits the name.

The new science of love indisputably demonstrates that we are united by nature; we are all imbued with the same existential fears and needs. Empathy is our birthright. We see it at the earliest age. One-year-olds, who do not as yet have language, will offer pats and hugs when a loved one cries "ouch" upon stubbing a toe. They will share food and toys with playmates and fetch articles and remove obstacles for adults, even at some cost to themselves.

De Waal argues that we should abandon the idea that humans are inherently selfish and only help others after mentally tallying up costs and benefits. The calculation has been made for us. We naturally favor empathy unless we are consumed by fear or rage. I have seen this with every couple I have worked with in the past thirty years. Once partners are able to let go of their desperate self-protection and engage emotionally, they respond to their mate's expressions of pain and vulnerability with compassion. This response continually confirms my belief in the basic goodness and generosity of human nature.

ALARM BELLS

When as a society we fail to grasp, honor, and nurture our need for emotional connection, we pay a huge price. Without loving attachment, we lapse into the morass of depression and anxiety that increasingly characterizes affluent Western cultures. In the United States, use of antianxiety drugs is up 30 percent from a decade ago, and one in five American adults takes at least one drug for anxiety or depression. The World Health Organization warns that children, too, are being medicated for depression to a disturbing degree. As John Bowlby noted, out-of-control anxiety and high levels of depression are natural consequences of emotional separation and disconnection.

Our bond with others is our ace in the hole when it comes to survival. It makes sense, then, as loneliness experts such as John Cacioppo suggest, that feelings of isolation and rejection are actually signals designed to move us to repair our social ties. We should heed them and refashion our individual and collective priorities. This means taking a hard look at the choices we make as individuals, as families, as citizens, and as active builders of our cities and civilizations.

On a personal level, perhaps we should think twice (or more) before running out for cosmetic procedures, especially concerning facial rejuvenation. The American Society of Plastic Surgeons notes that between 2011 and 2012, injections of fillers rose 5 percent. Shots of botulinum toxins, such as Botox, rose 8 percent, soaring past six million for the first time. Yes, these procedures can make us look younger, but so often they leave faces immobile and blank, wiped of laugh lines, furrowed brows, and all the other little signposts of emotion. How can we understand another's feelings when feelings aren't to be seen? Actress Julia Roberts, who tried Botox once, quickly swore off. "I was permanently surprised for a couple of months," she said. "It was not a cute look for me. My feeling is, I have three children who should know what emotion I'm feeling at the exact moment I'm feeling it...that is critical."

"It's no shock that we can't tell what the Botoxed are feeling," remarks David Neal, a psychologist at the University of Southern California. "But it turns out that people with frozen faces have little idea what we're feeling, either." In a recent experiment Neal and his colleagues asked women who were injected with fillers or Botox to look at photos of people's eyes and the area around them and match them with the names of positive or negative feelings. The Botoxed women were significantly less accurate in their assessments than women who had not been injected with the paralytic. The probable reason: their faces can't move to mimic the expressions they are seeing. Remember those mirror neurons! Botox not only deadens nerves, it deadens communication.

If we stay on the individual level, we could talk endlessly about many issues. Time, for instance. Over the last decades we have increased the hours we spend working, so much so that the line between work and personal time has all but disappeared. Prioritizing is all about assigning time. Discussions of why love dies often seem to miss the obvious: without time and attention all relationships evaporate. I wonder what happened to the idea of sacred (from the Latin, meaning "worthy of devotion") time that we set aside for our lovers and families? As I stroll the streets of Jerusalem on the Sabbath, the shops are closed and the people are walking to temple, talking with family and friends. In the city I live in, and in most cities in the world, we have decided that commerce and convenience come first: the malls and supermarkets have to be open, and Sunday is now like any other day.

When my children were young, I did no professional work on Sundays; they were reserved for family outings and time to be with my partner. This decision was much easier to make when collectively we reserved this one day out of seven for the things we considered holy. What happened to that? Having no space or time set aside for connection with loved ones and community now seems normal. Now we have to consciously decide to swim against the tide of nonstop multitasking to turn toward those we love. We, as a society, must not just leave this to individual judgment but begin to seriously examine the impact of our laws and broad social policies on the quality of our most important relationships and foster a social structure that actively promotes secure and lasting bonds.

We might begin by looking at the implications of business policies on families, especially at times of stress and transition. We know that relationship breakdown often begins with the birth of a child. If we take this as a warning, what might we do differently? On a family level, we'd do well to follow the lead of Norway, Sweden, and Denmark, which are already leaders in "bonding matters" policies. These countries offer between twelve and sixteen months of full paid leave to mothers and fathers, who can decide how to share that time between them. Canada offers almost a year off, but with much less remuneration. The United States has no national law mandating paid time off for new parents, although some states—California, for example—have begun to offer short leaves. Parental leave makes financial, social, and love sense from many angles. It promotes marriage stability, gives mothers who usually shoulder most child care responsibilities some respite, fosters bonding between mother and child, and also promotes infant health, thus getting baby off to a good emotional and physical start. Studies also indicate that the longer the leave a father takes,

the more engaged he is with his children and the better the youngsters do in developing mental and social skills. If governments
want to support the most basic building block of society — committed couples and their families — offering solid paid leave to
partners as they go through this critical transition makes excellent
sense.

We also seem intent on building our cities with no regard at
all for the social and bonding nature of the human animal. City
governments seem to have forgotten how to spell the word *community*. Despite visionaries, like the late Jane Jacobs, who have written
paeans to the merits of small, organic communities, where people know each other and live, work, celebrate, and survive in a
web of shared support, governments have been wedded to a "bigger is better" philosophy. This seems to have gathered momentum
in the 1960s and 1970s, when old neighborhoods were routinely
bulldozed to erect tall new buildings and create superhighways.
Now such a philosophy is the norm. While the intention may be
good, and old neighborhoods sometimes need razing, the effect
on a human and community level is often disastrous. People have
spiffier surroundings, but their close relationships with neighbors,
the people they see every day and count on for company and help,
are completely disrupted. Warehousing people is efficient, but it
defeats the human need for belonging and social connection.

This was brought home to me recently in New York City one
afternoon in Washington Square Park. All the dog owners brought
their pooches out to play, and they sat, chatting away, on the seats
in the big-dog and small-dog enclosures. I leaned over the fence
and spoke to Mildred as she kept an eye on Doodlebug, her over-
sexed Chihuahua. She told me she had lived in one of the big
apartment blocks near the park for thirty years. I commented, "Oh,
you must know the people in your building very well, then." She
looked at me, horrified. "I don't speak to them," she protested in

a high, squeaky voice. "I just bring Doodlebug here and chat to a few folks I know sometimes is all." I didn't know what to say. It seemed to me that the people on the benches were, in fact, keeping their distance, disturbed by Doodlebug's rampant sexuality, which extended even to beer cans and an elderly gentleman's left foot. But suddenly I felt sad. The park gave people a common place to walk, talk, and connect, but this apparently did not happen in the buildings these people called home.

How very different from my own upbringing! When I was growing up in a small English town, my life centered around my father's pub, with its 130 or so regular patrons (we seem designed to flourish in groups about this size; that's about the same number as in the tribes of hunter-gatherers that we originally evolved from). These patrons drank, flirted, caroused, bartered, worshipped, and cried together. The same people who appeared in the chorus of the Gilbert and Sullivan operetta at the town hall argued politics with my father and pinched my mother's elegant black-clad backside. It was a rich, wild, and almost disreputable environment for a little girl to grow up in, and at every minute I knew I was totally safe and cared for.

There are, of course, a small number of creative modern communities in North America, such as the award-winning Kentlands, outside of Washington, DC, that offer us models of modern habitation that honor our need for connection. As I drove into this town to visit a friend who recently moved from the big city, I thought I was entering an old village. Small parks and squares were dotted everywhere; shops, churches, and a theater popped up on corners. People sat on large porches and called to their neighbors or walked around the small lakes in the center of the town. This, I thought to myself, is made for humans; I knew I could live there. My friend in the town, Kathryn, tells me, "I know my neighbors, and I walk my dog with the same folks every morning. I know that if I need anything when my husband is away, there are people I can call on. The guys in the deli over there know the salami sandwiches I like. They make them for me without the olives. We like it here, and I feel so much calmer than when we lived downtown in DC."

Our love relationships bear a great weight if the only place we are seen and known and called to is in our own living rooms. We can build relationship-friendly environments made for *Homo sapiens* and *Homo vinculum* that do not ignore the imperatives of human connection. If we continue on our present path, dismissing the necessities of our nature, we will find ourselves ever more isolated and, as John Bowlby would put it, ever more in a state of emotional starvation.

THE TECHNOLOGY TRAP

Even as we understand how relational we truly are, the basic currency of social connection—face-to-face contact and simple conversation—is becoming marginalized. I was recently in a small neighborhood restaurant in Naples, Italy, and watched a large family claim a table that waiters had scurried to set up. At one end sat the elderly paterfamilias with his four sons and their wives; at the other, nine children. I settled in to watch the rich circus of Latin family life. And indeed, there was much laughing, hugging, arguing, and remonstrating—but only at the adult end of the table. The other end was totally silent. Eight of the children sat engrossed by the small electronic screens they held a few inches from their faces. Not for one moment did they ever speak or look at each other or at the adults, and they completely ignored the only child without an electronic device. Eventually this boy began to bellow in protest and was comforted by his mother, who turned his chair to face the adult group. In spite of the warm Mediterranean night, I felt chilled.

Pamela Eyring, director of the Protocol School of Washington, which teaches social manners to corporate and government clients, has identified four stages—confusion, discomfort, irritation, and, finally, outrage—of what she terms "BlackBerry abandonment": the feeling a person suffers when trying to connect with a devotee of the electronic gadget. She adds that since personal and business relationships rely on making others feel valued, devices such as iPhones put these relationships at risk. She calls an obsession with iPhones "cell-fishness." But this is about more than an issue of etiquette or a lack of consideration for others. A survey by the consumer electronics review site Retrevo.com found that 10 percent of people under the age of twenty-five don't see anything wrong with texting during sex!

Some say that all our electronic gadgetry is keeping us more connected. But while sharing information on a screen has its uses, it is a shallow connection, not the deep emotional engagement needed for any kind of meaningful relationship. Texting and e-mails are set up for volume, velocity, and multitasking—that is, the splitting of attention. They create an illusion of connection. The danger is that they also set up a new way of relating in which we are continually in touch but emotionally detached.

In her book *Alone Together,* Sherry Turkle, professor at MIT, suggests that, in the last fifteen years, our tools have begun to shape us and our connection with others, so that we now "expect more from technology and less from each other." Turkle analyzes detailed interviews with technology users and conducts formal studies on the impact of robots on them. Her studies examine a moving target. A 2010 Nielsen survey reported that the average teen sends more than three thousand text messages a month, and this surely will increase.

But this is just the beginning. Parents buy children interactive ZhuZhu robotic pet hamsters who are, according to the ads, "living to feel the love," or the more sophisticated robot puppy AIBO, from Sony. The adults in Turkle's book speak of interacting with AIBO in the beginning just for amusement, but admit to later turning to the robot when they are "lonely." Then there is Paro, the furry baby harp seal robot who holds eye contact and is designed to be "therapeutic" for depressed adults and others. These kinds of substitutions, Turkle argues, have "put the real on the run." Technology tends to reduce relationships to simple bytes, then the bytes become the accepted norm. To borrow a phrase from the late Daniel Moynihan, noted sociologist and U.S. senator, it's defining relationships down.

David Levy, in his book *Love and Sex with Robots,* suggests that soon love with robots will be as accepted as love with human beings. Roxxxy, the first sex robot, or "girlfriend," is leagues ahead of blow-up dolls. Introduced in 2010, Roxxxy can be customized in terms of her physical attributes and even her personality. And she has electronically warmed skin, internal organs that pulse, and can make programmed conversation—but only about sex or soccer! What more could someone wish for?!

The movement to have small robots take care of our most vulnerable citizens, children and the elderly, is also gaining momentum. Paro, for example, is being touted as the solution for lonely seniors. It understands about five hundred English words, and seniors seem to like it—unless, of course, they can choose to talk to a real person.

To me, robots like Roxxxy and Paro reflect our growing sense of failure and resignation where close relationships are concerned as well as our profound lack of awareness about our need for intimate emotional connection. The one thing that robots cannot do is feel emotion; they offer a counterfeit performance that imitates connection. Just as distressed couples do, when we become lost and desperate, we pick up solutions that seem to offer immediate comfort but further distort our ability to really connect with another person. In a lonely society, a substitute relationship may be better than none at all, but substitution blurs into replacement, and replacement becomes preferred.

Howard, one of the people interviewed by Sherry Turkle about the "artificial companionship" offered by a robot, comments, "Well, people are risky...robots are safe." People like Howard seem to relate to their robots as though they are sentient and emotional, even though they say that they "know" the robot is a machine. They "attach to" these machines, which mimic listening and concern, unable to resist the idea that the machine "cares." As Turkle points out, our desire for caring is so absolute that it takes precedence over our knowledge of the machine's indifference. Substitute pseudo-attachments can be seductive, but in the end they take us farther and farther away from the real thing—a loving, felt sense of connection that requires moments of full, absorbing attention and a tuning in to the nuances of emotion.

What we ask of robots shows us what is missing in our lives. As we turn to technology instead of each other, face-to-face connection is lessening and real ties are weakening. Turkle concludes that "a machine taken as a friend demeans what we mean by friendship; after all, we don't count on our cyber friends to come by when we are ill or console us in the face of loss." Indeed, we are wired to count not just on our friends but also on our loved ones to do that.

When I listen to couples describing how they spend their time, it hits me that tapping on the iPad and the computer and watching TV's ironically termed "reality" shows diminish our opportunities to engage with and care for another person. Technology in general, just like pornography, offers us lousy models for connecting and bonding with other people. We become accustomed to the simplified, the superficial, and the sensational; we turn to the endless stories of celebrity relationships rather than learning to craft our own. As political scientist Robert Putnam notes in his seminal book on the loss of social connection in Western societies, *Bowling Alone,* "Good socialization is a prerequisite for life online, not an effect of it: without a real world counterpart, Internet contact gets ranty, dishonest and weird." We communicate more and more and say less and less. In a good love relationship, if we can turn off the screen, we can learn to say what really matters to us in ways that build connection.

There is a chicken-and-egg factor here. Isolation, I am arguing, is an effect of our obsession with technology, but growing social isolation also creates this obsession. More than at any time in human history, we live alone. In 1950, only four million folks in the United States lived on their own; in 2012, more than thirty million did. That's 28 percent of households—the same percentage as in Canada; in the UK., it's 34 percent. Eric Klinenberg, a sociologist at New York University and the author of *Going Solo: The Extraordinary Rise and Surprising Appeal of Living Alone,* observes that these skyrocketing statistics tell us that "a remarkable social experiment" is occurring.

How does this momentous shift fit into the design of the creature we call a human being? Technology is being touted as the solution to our growing isolation, but in fact it is just part of the problem. Real connection with others is being crowded out by virtual kinship. This is worrying even on just a pragmatic level. Psychologists point out that cooperation, on which society depends, is a learned skill that until recently almost everyone acquired. Today, however, fewer and fewer people have the ability to collaborate; instead they withdraw from group tasks and social life.

My client Marjorie stares at the floor in my office while she tells me, "I am on my own since my marriage split up and my son left. So it's quiet, and I just have to deal with me. I have gotten used to it. I watch TV a lot and play video games. But at work, there is this buzz. Lots of people, and they want me to listen and help them do stuff. I get overwhelmed and irritated. And now my supervisor says she is going to lay me off. I told her that was fine. But it isn't, really. Then I really would be all by myself—and broke."

ONLINE "AFFAIRS"

You do not have to go far to find other examples of just how dismissive of relationships Western societies have become or to grasp how relationships are increasingly viewed as commodities. Standing at my kitchen counter at breakfast time one day, near the end of writing this book, I casually began to read an article in my local newspaper. Noel Biderman, CEO of Ashley Madison, an online dating service for married folks who want to cheat, had just announced the results of the company's survey of its clientele. I scanned the findings on the unfaithful: the typical man is in his forties, the woman is thirty-one; they have been married for about five years and commonly have a daughter around age two or three; and they are affluent. Interesting. And sad.

But the next line really got my attention. In fact, I dropped my toast! Ottawa, the small, sleepy capital city where I spend much of my time, is a hot spot for customers buying the opportunity to have an affair, having the most paid subscribers to Ashley Madison per capita. And my very own neighborhood, bounded by a lazy river and quiet canal, full of old houses and huge trees, coffee shops and florists, is the very hottest local area for customers. I won't tell you what I did next, but it was loud. My dog took a dive under the table. Maybe it's capital cities: the top subscriber locale in the United States is Washington, DC.

According to Biderman, the most popular time for women who have kids to sign up is right after Mother's Day! These moms say they get little attention from their partner and miss feeling emotionally supported and sexually desired. I felt sad and a bit sick that so many of my neighbors might feel that the only place they can turn to to find a way out of this kind of distress is a website that takes advantage of their vulnerability and has them pay for the privilege of further injuring their relationship.

Later in the morning, I got a long e-mail from a colleague who was writing to me about relationship education programs. His point was that there are a number of solid programs that seem to help couples improve their relationship before irrevocable harm has been done, but that enrollment in these programs tends to be low. In general, relationship education is a hard sell. Then I remembered an article indicating that the majority of people who sign up for sites that facilitate cheating never, in fact, cheat. They join to *flirt* with cheating, to *talk* with someone about cheating, but rarely meet up with a potential lover. In all likelihood, they are feeling abandoned and alone and looking for a distraction, a fantasy that promises relief from their pain and makes them feel as if they have options.

I had an overwhelming desire to run out into the streets and yell to all my neighbors, "Listen up. You just shut down those dead-end quick-and-easy lover sites. They won't help. Go find yourself a program to help you and your partner reconnect." Why don't people do that? Maybe they do not know how to even begin to talk to their partner about their distress, or they believe their partner would not agree to attend such a program. Perhaps they find going to a program scarier than conducting a fantasy affair on the Internet.

But when I think of all the unhappy couples I have seen, and that so many of them wait years before seeking a couple therapist, I am convinced that resorting to the Web is about a collective sense that love is something that just happens to us, that we have no control over its vagaries, and that our only recourse when it turns against us is to seek distance and distraction.

MAKING A BETTER LIFE AND WORLD
WITH LOVE SENSE

If we take the research on the power of loving bonds seriously, then what? How can we begin to use the lessons of the new science? We can answer this question in two ways. We can improve our individual decisions and practices with our loved ones, and we can try to actively shape society to recognize, honor, and prioritize our innate need for connection.

On a personal level, if we consider that in a relationship the connection between two partners is always variable, oscillating between moments of attunement and synchrony and moments of misattunement and disconnection, then we can set up rituals in which we intentionally reset the dial, reattune, and reconnect. I remember a client, Charlene, telling me of a game she and her small son engaged in that morphed into a reattunement ritual with her husband. It was called Where Are You? With her son, it was a hide-and-seek game, but with her spouse it was about moving into being emotionally present. "Where are you now?" she would ask. He had learned to report his emotional state at that moment. "I am feeling fried. Too much to do. I like that you ask me this, though. It calms me down." Then she would reciprocate by telling him what she was feeling. This simple routine stopped days rolling by, said Charlene, with "no real personal connection, no one tuning in to and opening the emotional channel, if you know what I mean."

Anniversaries are a great opportunity for reconnection. What if we renewed our wedding vows not after ten or twenty years together, as some of us do, but every year? We could talk with our partner to review what has changed in the past year, what the delights and disappointments have been, and also what the theme of our own love story has been over the long term. Then, using our original vows as a guide, we could make those vows again with a focus on the following year and the ways in which we intend to nurture our love for each other. With understanding, we can learn to be more attentive and deliberate about our most precious and necessary relationships.

On a societal level, the most obvious implication of the new science of bonding is that we must educate for connection. The most organic way to do this is to support couples in their efforts to create loving bonds and be responsive parents. We should acknowledge, as Frans de Waal notes, that "there is no escaping the reality that we are dependent on others. It is a given. If dependency/vulnerability is recognized and handled well in loving relationships...it is the source of the best human qualities, empathy, kindness and cooperation." We need to educate for qualities such as empathy, which is at least as relevant to health, happiness, and citizenship as arithmetic. But do we know *how* to teach these qualities?

CATCHING EMPATHY

Empathy can be "caught, not taught," says educator Mary Gordon, who founded the program Roots of Empathy (ROE). Nine times a year, she brings one particular mother and her maturing infant to an elementary school to coach children in grades 1 through 8 in "emotional literacy" and give them "the picture of what love looks like." Before each session, an instructor explains the language of feelings and attachment and gives tips on observing the mother and infant interacting. After the visit, the children review the visit and discuss their own experiences and feelings of, say, fear or frustration and the ways in which they can deal with or help others who have these emotions.

An ROE instructor might ask, "What is the baby trying to tell us right now?" "How did the baby tell the mother that she needed her?" "What did the mother do?" "What can the baby do now that she couldn't do last time?" "What should we do?" The children also receive art, drama, and writing assignments to further explore feelings of attachment as well as gain basic information on human development.

By end of 2012, 450,000 children in Canada and Australia had gone through the program. Gordon believes that far too many people are "emotional islands, cut off from meaningful connection to others because they can't speak the universal language of their emotions." Research conducted since 2000 by professor of education Kimberly Schonert-Reichl of the University of British Columbia finds that emotional understanding and pro-social behaviors increase, and aggressive behavior decreases, in children who receive ROE instruction. Nasty behavior dropped 61 percent in ROE children (compared with an increase of 67 percent in youngsters who weren't in the program). ROE children generally became more cooperative, helpful, and kinder, and peers rated them so. For example, when children in an ROE class explained to schoolmates how humiliated and unhappy a nine-year-old who is wheelchair-bound and drools uncontrollably feels when students call him names and mock him, the harassment stopped.

The program creates a safe environment where children can discuss their feelings and learn to regulate them. The implications are broad. Programs like this could help curb bullying, which has reached epidemic proportions in North American schools and is a strong predictor of delinquency as well as later criminal activity, alcoholism, and mental health problems. Today, one in five children and adolescents experience psychological difficulties, including depression and anxiety, severe enough to warrant treatment.

Some teachers call ROE the fourth R (after reading, 'riting, and 'rithmetic) as a way of stating how much they value this program and how it builds essential basic social skills. Indeed, this kind of program does seem to help boost basic abilities and also general academic achievement. Research shows that the social skills a child exhibits in grade 3 more strongly predict academic success in grade 8 than does academic prowess. "Too many children are unable to learn because they are in so much social pain," says Gordon. Their energy is spent on being vigilant for threats and managing fear; there's little left over for studying.

By increasing emotional engagement and responsiveness, Gordon hopes to build children who will be better friends, better citizens, and eventually become better parents themselves. We now have "so many people running on empty," she says, and draws an analogy. "There is fluoride in our water supply to prevent tooth decay . . . we need empathy in our water to prevent social decay."

The real question is, how much do we value human connection? It seems we are ambivalent. In 2009, the British Columbia provincial government cut the funding for the Roots of Empathy program, but in 2012, a new government restored enough funding for it to be reinstated in 360 classrooms. The question of whether we value human connection and empathy enough to educate for it and deliberately promote it perhaps depends on how we see civilization itself. When he came to England in the 1930s for talks on Indian self-rule, Mahatma Gandhi was asked by a reporter what he thought of Western civilization. Gandhi replied, "I think it would be a very good idea."

The word *civilization* comes from the Latin word for "citizens" and signifies an advanced state of human social development and organization. Do we judge this state by how high our buildings are and how many fancy goods are in our stores? Or do we judge it by the quality of our relationships?

In November of 2012, walking in the streets of old Jerusalem, I saw, in this long-divided and fractious city, two little girls of around three years old walking calmly hand in hand down the narrow stone street. After seeing young Israeli soldiers cradling their automatic weapons at every city gate, this innocent image of connection and assumed safety felt reassuring. In Denmark, people leave their children unattended outside, trusting that no one will take them or hurt them. In Oslo, there are no beggars on the streets; if someone begins to beg, people come up and offer to help or call a city staffer for assistance. In small-town America and Canada, where communities are more closely knit than they are in cities, doors are left open and cars unlocked. Civilization, it seems to me, works most effectively when we take our social capital seriously and cultivate it.

LEADERSHIP AND CONNECTION

As we understand relationships and how they foster growth and strength in us as individuals, we can extend this understanding to the world of work and use it to create more effective leaders. Scientific descriptions of successful leaders in business and the armed forces reflect the qualities of model attachment figures. They are tuned in and responsive to subordinates, they give them guidance, offer them challenges, support their initiatives, and foster their self-confidence and self-worth. A good example is found in the movie *Saving Private Ryan*. Captain Miller, played by Tom Hanks, teaches his men to trust him and each other; he turns them into a powerful, cohesive team that can accomplish its mission. Trust is the glue that turns a group of men into a unit, just as it is the adhesive that turns two individuals into a bonded pair.

This holds true in sports as well. Psychologist Michael Kraus and his colleagues at the University of California, Berkeley, found that the best predictor of which National Basketball Association team was going to win the final playoffs in the 2008–9 season was not early-season performance but the number of times team members reached for and touched each other in the first game. Reassuring touch from teammates appears to enhance the sense that they can rely on each other, increases cooperation, and frees players to focus completely on the game. Human connection works!

Mario Mikulincer, of the Interdisciplinary Center in Herzliya, has examined the link between attachment and leadership in the Israeli army. In one study, he identified the attachment styles of young military recruits at the start of four months of intensive training. When they finished, he asked them to name those who should become leaders. The recruits' nominees all had secure attachment styles.

In another study, Mikulincer asked 200 people, including officers in the Israel Defense Forces and business managers in the public and private sectors, to fill out questionnaires about their attachment style and their motives for seeking leadership positions. More avoidant leaders generally endorsed statements that demonstrated strength, toughness, independence, dictatorial decision making, and staying removed from followers; their attitude is summed up by the phrase "I like the pleasure of having control over people." More anxiously attached leaders tended to check statements that showed a desire to cultivate subordinates' growth ("I devote effort to the personal development of my followers"). But they were so unsure of themselves that their followers reported being uncertain as to exactly how they could effectively contribute to the unit's performance in key tasks and problem solving. Anxiety interferes with effective communication.

In a third study, Mikulincer surveyed soldiers about group unity ("Does the team work well together?" "Does your group have a high level of consensus?"). Soldiers rated both anxious and avoidant leaders as poor at building group cohesion, but in different ways. Anxiously attached officers were judged as deficient in providing direction in task-oriented situations, while avoidant leaders were deficient in emotion-focused duties, such as building morale.

Avoidant leaders received particularly poor marks when soldiers were questioned after stressful combat training. The more avoidant the officer, the less he was viewed by his unit as a supportive influence. Indeed, soldiers, even those who rated themselves before training as very secure, reported feeling nervous strain and becoming depressed.

Mikulincer and his team conclude that avoidant leaders are inclined to dismiss their own and others' emotions and to hold a negative general attitude toward others. They demoralize followers, reducing enthusiasm for group tasks. Anxious leaders, on the other hand, doubt their own abilities and communicate uncertainty to subordinates, and that makes the team hesitant to act and lowers productivity. This kind of research expands our concepts of effective leadership. It demonstrates that in leadership, as in other spheres of life, it is those who can manage their emotions and connect well with others who are most able to create the secure structure that promotes high achievement.

GOOD CITIZENS

For many years the goals of personal growth and autonomy have been seen as somehow antithetical to bonding and our need for others. In fact, secure connection is the fertile soil out of which confident, resilient, and independent human beings spring. This sense of safe connection and openness sets us up for what noted psychologist Abraham Maslow in the 1970s called self-actualization. More secure folks tend to accept who they are and see themselves and others as deserving of caring and concern. When we are securely attached we have a more positive, balanced, complex, and coherent sense of self. In fact, securely attached people show fewer discrepancies between their stated actual and ideal traits. At the end of therapy, Anita tells me, "When I am closer to Ken, I feel more confident, and just better about myself. Knowing I am special to him helps me accept my fears and know that I can deal with them. It's okay to be afraid sometimes. I don't have to put on this facade of coolness anymore. It's simpler in the end to just recognize your vulnerabilities—that is the best way to be."

But more than this, since securely attached people live in what they perceive as a safe world, they are less self-absorbed and less preoccupied with threats than are anxious or avoidant people. This enables them to focus on, empathize with, and be tolerant of others. John Bowlby believed that when given loving care as children, human beings are naturally empathetic and altruistic. He also believed that insecure attachment tends to suppress or override our natural tendency to care for others. And it seems that secure connection with others does indeed further in us the ability to respond compassionately to their needs.

Psychologists can now "prime," or turn on, a sense of security in a lab and, for at least a short time, expand a person's compassion. Mikulincer and his colleagues have been examining how a taste of loving connection affects our ability to feel for and act in the interest of others. In one study, students read either a story of a person providing loving care and support to someone in distress or a story describing a person who voiced the sort of platitudes expected to lift a general listener's mood. After that, they read a story about a student whose parents had just been killed in a car accident. Then they were told to rate how much sympathy and compassion as well as personal sorrow they felt for the student. After the attachment-priming story, everyone felt more compassion for the bereaved student than they did after the story containing good-mood phrases. But avoidant and anxious readers felt less than did securely attached readers. And the anxiously attached also reported being more personally upset by the tale than those with other attachment styles. It became about their sadness, not the student's.

If we become more connected, do higher levels of empathy really translate into action—into a willingness to help a person in distress? In another experiment, subjects were asked to identify the people they were closest to (through questions such as "Who is the person you turn to when you are feeling down?"). Most had three attachment figures. Then the subjects were seated at a computer and told to look at strings of letters that appeared on the screen and determine if they constituted a word. Buried in each string was a rapid subliminal (that is, subconscious) presentation of the name of one of their loved ones. Another group of folks also scrolled through letters but were told to deliberately (that is, consciously) think of a significant positive incident with a particular loved one. A comparison group went through the same process, but the name provided in the experiment was of someone who was an acquaintance, not a loved one.

All the subjects then were asked, ostensibly as part of a different experiment, to watch a woman in another lab performing a series of increasingly aversive tasks. As they watched, the researcher in the lab insisted very strongly that the tasks had to be done or the whole project would be ruined. In fact, subjects were watching an actress in a pre-recorded video clip. The grim tasks progressed from looking at photos of a gory accident to sinking a hand into frigid water to holding a crawling tarantula or squirming rat. As the video progressed, the woman's agony intensified, and at one point she pleaded for someone to replace her. Viewers were asked to rate their own distress and also their willingness to take her place.

This study was replicated five times with five different groups of participants. Each time, subjects who were primed, consciously and unconsciously, with the name of a loved one reported being more upset by the woman's predicament and feeling more compassion for her than were subjects primed with an acquaintance's name. They were also more likely to offer to take her place.

Within the attachment-primed group, there were some subtle differences. Avoidant folks rated themselves as feeling less compassionate and alarmed by the woman's situation and less willing to take her place than did secure observers. Anxious folks became very distressed by her predicament but were also less willing to take her place than were secures. Priming our attachment system appears to turn on our altruistic, care-giving system for a moment, but we seem to need a certain level of security before we move into compassionate action. Avoidants appear to maintain a certain detachment, while the anxious are gripped by their own suffering.

Security also creates a tolerant attitude toward what is unfamiliar or novel. When you believe that others have your back, the unfamiliar is less threatening. Secure folks have more stable self-esteem and are less inclined to denigrate others (as in "We—my group and I—are better than they are"). So it makes sense that this kind of priming security research has also been used to test whether it is possible to promote tolerance between groups of people who are different and between whom there may be enmity, such as homosexuals and heterosexuals or Arabs and Israelis.

When research subjects were asked to visualize the face of a loving attachment figure immediately before rating their reaction to images of a member of a "foreign" group, their previously assessed negative attitude toward these out-groups disappeared entirely. Tuning in to their sense of safe connection made them less judgmental. Where before they had described out-group members as "sleazy, spineless, and lazy," they subsequently regarded them as "trustworthy, warm, and kind"—just like members of their own group. This positive view was present even when a sense of threat was induced, as when subjects were told that members of their own group had recently been insulted by someone from the "foreign" group.

But can turning on a felt sense of secure attachment actually reduce aggression between warring groups? This is a little hard to test in a lab. Mikulincer decided to try the Hot Sauce study. This is the research version of a common childhood ruse: offering a kid you're on the outs with what appears to be a gummy worm but is actually a real worm. Do you watch with glee as he eats it and begins to heave (or not)? In studies of aggression, the question is: how much hot sauce is a subject willing to push on someone else?

One such study involved a group of Israeli students. On several occasions, each student was repeatedly subliminally primed for 20 milliseconds with the name of a main attachment figure, that of a friend who was not a security figure, and that of an acquaintance. After each priming, the subjects were asked to give an Arab and an Israeli a quantity of hot sauce to sample; the students were told that both very much disliked spicy foods. Those who had been primed for attachment gave small equal amounts of sauce to both the Arab and Israeli. When primed with the names of the other two people, however, the subjects repeatedly gave larger amounts of hot sauce to the Arab than they did to the Israeli.

The implications of this kind of research for society are obvious. Secure relationships with parents and partners make for more compassionate and caring citizens, who will be more tolerant of those who are not like them. These priming procedures remind me of religious rituals in which people are encouraged to become more compassionate by visualizing the loving Buddhist goddess Tara, praying to the loving Christian God, or simply meditating with gratitude on the benevolence of the universe that surrounds them.

The findings of priming studies like those above suggest that our feeling for and willingness to act on behalf of others has plasticity; it can be shaped, especially if images that awaken our deepest need to belong and be held in loving connection are evoked. Understanding attachment shows us how loving parenting and partnering translates into a kinder, more humane society. You may not realize it, but when you hold your child or respond to your partner's call, you are sculpting a civilization.

A NEW SOCIETY

In terms of revolutions, the latter half of the 18th century was hot! First there was the American Revolution of 1775, and then the French Revolution of 1789. Both these momentous uprisings enshrined ideals that are reflected in the founding documents of many modern democracies. The American War of Independence hallowed the rights of the individual to liberty and equality. The French Revolution consecrated an additional principle: the brotherhood of man. *Liberté, égalité, fraternité* is the French national motto.

There are some people, such as the late Charles Gonthier, a justice of the Supreme Court of Canada, who suggest that we have forgotten the core values of empathy, trust, and commitment that characterize this last pillar of democracy—fraternity. The new science of human bonding extends this last element beyond a recognition of the need for fellow feeling and cooperation between neighbors. It places a recognition of our emotional and physical interdependence and the need for safe, trusting, caring relationships at the very core of human nature and of a truly human society. This new science is not just a formula for romantic and family love. It is a blueprint for the reform and optimal development of human society.

We could start by considering ways to raise awareness of the perils of loneliness and by validating our need for belonging and support. Can you imagine if we took a small amount of the $641 million spent on antismoking campaigns in the United States in 2010 and created an anti-disconnection campaign, publicizing the dangers of emotional isolation? Emotional isolation has actually

293

been found to be more dangerous to our health than smoking, so this suggestion has much logic to it. We could put banners up in our cities asking people, "Who did you reach out to today?" or even telling them, "Take someone you know (or don't know!) for a latte today. It's good for your health."

There are a thousand ways we could bring building relationships to front and center. Let's just make a very short list. We could write letters to our government representatives pressing them to help create more connection-friendly communities with public spaces that promote easy social engagement. We could also ask our representatives to widely promote basic relationship education for partners and for parents. We could encourage our radio and TV stations to offer serious and informative programs on relationship issues. We could revise professional education programs to inform doctors and psychologists of research that shows that including a patient's partner increases the efficacy of treatment for everything from anxiety to heart disease.

The new awareness of how relationships affect our health and well-being has already resulted in small, specific changes that we can, I hope, expand and consolidate. Some of them, such as the Roots of Empathy program, are mentioned here. Some communities are contributing to public relationship education with small initiatives—for example, religious groups are recruiting happy middle-aged and older couples to act as group consultants to younger couples by sharing their experience of long-term love. Most cities have distress lines operated by volunteers, and some agencies offer free training to young people who are interested in staffing hotlines for those who need and want someone to confide in. Some of our youth already take a gap year between high school and university and choose programs dedicated to service to others. This practice started in the U.K. in the 1960s, and programs like AmeriCorps, created in the 1990s by U.S. president Bill Clinton, now promote youngsters' involvement in projects such as community education and environmental cleanup. This initiative could be expanded so that, as part of a humanistic education, all students would be required to dedicate a year to direct community service as a prerequisite for entering university or college.

In "Forty-Four Juvenile Thieves," John Bowlby explained that the defiance, desperation, and rage typically found in young delinquents is largely a reflection of family dysfunction and the resulting disconnection from others. Optimal family functioning starts and ends with the bond between parents; it is infinitely harder to pull off responsive parenting without such a bond, especially without the web of support that so many small and intimate communities used to provide. Simple indicators of connection tell us how far off track we are. Surveys show that U.S. families now rarely share meals together and that parents spend little time talking with their offspring. Almost half of American two-year-olds watch TV for at least three hours each day. A 2007 UNICEF report on child well-being in the world's twenty-one richest nations rated the United States, with its chaotic family structures, troubled family relationships, and exposure to violence, second worst. Fragile, unstable families do not bode well for the creation of strong, safe emotional relationships that will stabilize children and help them grow into well-functioning adults and citizens of the world.

In addition to awakening to the fact that we need to take care of the planet we live on, we must also recognize that we need to guard the ecological niche we occupy—close connection with others. Our overwhelmed health systems are already passing the buck back to families by releasing people from hospitals sooner and expecting family members to care for the sick and also the elderly. But increasingly, caregivers are overwhelmed and unable to cope with the burden. The social problems that our society faces can only be effectively addressed by strengthening adult love relationships, families, and communities.

The quality of love relationships is no longer simply a personal affair. When these relationships blossom, we all benefit; when they become distressed or break, we all suffer. Dissolving a marriage costs individuals and also taxpayers. In the United States, it is estimated that on average, each divorce drains thirty thousand dollars out of the public treasury; the money is spent on food, housing, and health care for needy single-parent families as well as child-support enforcement. There are indirect costs as well: physical and mental health problems, lost work time, addiction, and crime.

In light of the above, governments around the world are starting to offer supportive services to couples in serious difficulty or facing stressful transitions—for example, preparing for marriage or the empty nest. Relate, in Great Britain, and the National Marriage Coalition, in Australia, provide free information, advice, consultation, and referrals to professionals through the Internet. In the United States, the federal government has been giving small grants to local and religious organizations as part of its Healthy Marriage Initiative program. The results of these efforts have been inconsistent, possibly because most programs have not focused on the attachment issues that are at the root of relationship discord. Most couples do not need to learn how to call time-outs during fights; they need to understand the fundamentals of love and learn how to reach for each other and be responsive to each other's needs. I believe that my new educational program, Hold Me Tight: Conversations for Connection, based on the science described in this book, will bring better results.

The biggest hurdle, however, is not the design of education programs but getting couples to attend them. Many see such courses as an admission of failure. Jenny, a client in individual therapy for depression, tells me, "Well, I know my relationship with Russell sometimes sparks my feeling depressed, but we would never consider doing a relationship education program, let alone anything like couple therapy! After all, that is for couples who are divorcing or something. And it's private stuff. I don't know anyone who has done that. You can't really change love, can you? 'It works or it doesn't' is what my mother says. And my friends are convinced you can't hold on to that love feeling forever."

Jenny is telling me that intentionally working at understanding love does not seem natural or feasible. She has no inkling that love is something that can be actively shaped and controlled and that she can learn how to do it. I imagine if you had advocated parent training just a few decades ago, you would have received a similar response from most people. Largely because of John Bowlby's work on mother-and-child relationships, the culture of parenting has changed. There is now an endless stream of books, courses, websites, media articles, and parenting groups reflecting and shaping a new consciousness of what it means to be a parent and what a child needs. The greatest promise of the new science of love is simply this: it will create a similar new empowering consciousness about what it is to be a lover.

As we educate for love and begin to see romantic love as intelligible and malleable, we will be able to shift from an obsession with the "fall" part of love to the "make" aspect of love (which will mean more than just sex). We will develop more confidence in our ability to work with and mold our love relationships. The more you believe you can influence what happens in a marriage, the harder you will try to keep it and mend it. And the more commitment you show, the more effective your efforts will be and, ultimately, the more stable your relationship will become. This increased stability is the second great promise of the new science of love. We can make love last because we now know how to repair and renew it.

But awakened awareness and education will never be enough in and of itself. Our political thinking has a long way to go to catch up with our new understanding of human bonding and the power of secure connection. If we truly want to support safe-haven and secure relationships for adults and children, then governments and corporations also need to offer a wide range of supportive work-place policies, including paid leave for new mothers and fathers as well as for employees caring for sick children and elderly or infirm adults. Opponents argue that such policies are costly and undermine productivity and competitiveness. Evidence suggests the contrary. Studies of highly effective companies find that family-friendly policies pay off in *reduced* costs and *increased* productivity. Employees are more engaged and creative and stay with their companies longer. And clients are more satisfied, too.

Surveys find the greatest reported well-being and happiness not in the wealthiest nations but in those with the highest level of trust among citizens and the most bonding-friendly social policies. In fact, wealth seems to come with a high price tag; many studies show that becoming preoccupied with materialistic concerns goes along with a loss of empathy for and trust in others. The preoccupation with acquiring more possessions or reaching for greater highs from drugs or alcohol will not work as a substitute for connection with others. The need for emotional connection is so intrinsic to who we are that there can be no substitutes. If we acknowledge love sense, we can move forward. We can move toward a time when "true" love, being known, becomes simpler and easier and more accessible to all of us.

There is much talk across cultures of the time we live in as a tipping point for mankind. Shamans and holy men have seen this period as the beginning of the end of the world—or the beginning of a new cycle. The Mayans predicted destruction. The Incas and Tibetan Bon shamans foretold renewal and transformation. The Hopi prophesied a "time of turning the earth over."

Commentators in our modern world are noting that there seems to be a dramatic shift in awareness, a new empathetic consciousness developing. This would make sense, given that we are beginning to realize how very interdependent we all are on this small blue planet and how very easy it would be for us to destroy ourselves. We are beginning to realize, after the long cycles of evolution that got us to this point, that mankind's survival depends on our finding a way to connect and cooperate in our personal, communal, and political lives. The serious study of our most formative love relationships fits very well with this tipping-point perspective. And it definitely comes down on the side of optimism. We are on the cusp of glorious new discoveries in medicine and in physics that will make for a quantum leap in the evolution of our species. Learning to love and be loved has to be at the heart of this leap as well.

Philosopher Kwame Anthony Appiah of Princeton University makes the point that, "In life, the challenge is not so much to figure out how best to play the game; the challenge is to figure out what game you are playing." The science summarized in this book has the potential to change our game. In my opinion, the only game worth playing is that of building a more humane society, a society that fits with our core nature as social and bonding animals and offers us a real chance to find secure, lasting love relationships—those that allow us not only to survive but also to become fully and optimally alive. As the old song suggests, when we truly love, we love someone "body and soul." The word *soul* comes from Old English and means "vital breath." And we are never so alive as when we love.

The development of love sense offers us a way forward into a different kind of world, a world in which we honor our deep desire to belong, where we have a felt sense of connection to our own soul and that of others. Secure love calms and restores balance and equilibrium. In 2006, while visiting Vancouver, Canada, the Dalai Lama told his audience, "I am now seventy-one years old. I feel, still, deep in my mind, my first experience, my mother's care. I can still feel it. That immediately gives me inner peace, inner calmness." Secure love promotes exploration and growth, expanding inner and outer worlds. It allows for a world based on trust and touches the most human quality of all, the one that we all share, our vulnerability.

There is an old hymn, "Abide with Me," that touches something very deep inside me whenever I hear it sung. Though a prayer to God, to me this is a song of attachment. Every scientist mentioned in this book and most of you, too, will know why it makes me weep.

> *Abide with me; fast falls the eventide;*
> *The darkness deepens; still with me abide.*
> *When other helpers fail and comforts flee,*
> *Help of the helpless, O abide with me.*

Acknowledgments

I dedicate this book on the science of close relationships to the people who have taught me the most about this topic.

The first is my father, Arthur Driver, an English sailor and small-town publican, whose face came alive with delight every time his mouthy little working-class daughter challenged or disagreed with him. My mother, Winnifred, was the one who taught me that the only thing that mattered was courage, including the courage to reach out for and to others. My diminutive grandmother, Ethel, showed me that when you have a treasured loved one by your side, even the hard times in life can be full of joy. And my lifelong friend Father Anthony Storey, even though he believed in a different God than I did, taught me that holiness, which comes from the Old English word for "whole," is always about compassion and care for others.

I have learned about all the topics in this book—connection, disconnection, emotion, and bonding—from my family: my three children, Sarah, Tim, and Emma, and my amazing life partner, John Palmer Douglas. It is a life's work to study love and connection and then to actually try to live what you have learned, and my family has been most patient with my attempts to do just this.

Over the years, I have also learned from my inspiring clients—this book would not have been possible without them—and from my wonderful students at the University of Ottawa and at Alliant International University in San Diego. In addition, I have learned from my incredible group of colleagues, who travel all over the world with me and teach what we have learned about creating secure bonds to mental health professionals and couples. Special colleagues in the couple therapy field, such as John and Julie Gottman, have encouraged and supported me. As the study of bonding has grown, others from fields outside clinical psychology, such as neuroscientist Jim Coan and social psychologists Mario Mikulincer and Phil Shaver, have challenged and nurtured me. My dear colleagues at the International Centre for Excellence in Emotionally Focused Therapy (ICEEFT) and all its thirty or so affiliated centers have created a community of committed clinicians who offer me a professional family where uncertainties can be explored and discoveries expanded.

I must thank Tracy Behar, my editor at Little, Brown, for her patience and for her commitment to helping me write a second book on the topic of love relationships; additional thanks go to my always upbeat and dedicated agent, Miriam Altshuler. This book would not be possible without the astute analytical mind and rigorous editing skills of Anastasia Toufexis, who insisted that this somewhat academic treatise be relevant and even readable.

Finally, I dedicate this book to all those who struggle to understand what romantic love is and who, even when lost in moments of deep confusion and despair, again and again turn back to their loved ones and try to put their feet on the path that leads to secure connection. There are so many of us.

Resources

For more information about Dr. Sue Johnson, her books, and her work, go to www.drsuejohnson.com.

Dr. Johnson's book *Hold Me Tight: Seven Conversations for a Lifetime of Love* offers an introduction to and a lay version of Emotionally Focused Therapy (EFT), the couple therapy she developed. It is available from booksellers and at www.HoldMeTight.com. Also at this website is information on *Hold Me Tight: Conversations for Connection,* a DVD showing three couples as they change their relationship with these conversations. There is also a list of Hold Me Tight retreats for couples held throughout North America.

To find a therapist trained in EFT, go to www.ICEEFT.com, the website of the International Centre for Excellence in EFT. This nonprofit organization is dedicated to the practice and evaluation of cutting-edge couple and family interventions, the education of professional therapists, and basic research into the process of change in love relationships.

References

General

Blum, Deborah. *Love at Goon Park: Harry Harlow and the science of affection*. Perseus Publishing, 2002.

Bowlby, John. *A Secure Base: Clinical applications of attachment theory*. Basic Books, 1988.

Cacioppo, John, and William Patrick. *Loneliness: Human nature and the need for social connection*. Norton, 2008.

Cassidy, Jude, and Phillip R. Shaver, editors. *Handbook of Attachment: Theory, research, and clinical implications*. Guilford Press, 2008.

Coontz, Stephanie. *Marriage, A History: From obedience to intimacy or how love conquered marriage*. Viking, 2005.

Cozolino, Louis. *The Neuroscience of Human Relationships: Attachment and the developing social brain*. Norton, 2006.

de Waal, Frans. *The Age of Empathy: Nature's lessons for a kinder society*. Harmony Books, 2009.

Ekman, Paul. *Emotions Revealed*. Henry Holt, 2003.

Fine, Cordelia. *Delusions of Gender: How our minds, society, and neurosexism create difference*. Norton, 2010.

Goleman, Daniel. *Social Intelligence: The new science of human relationships.* Bantam, 2006.

Gottman, John. *The Seven Principles for Making Marriage Work.* Crown, 1999.

Iacoboni, Marco. *Mirroring People: The new science of how we connect with others.* Farrar, Straus and Giroux, 2008.

Johnson, Susan M. *The Practice of Emotionally Focused Couple Therapy: Creating connection.* Brunner/Routledge, 2004.

Karen, Robert. *Becoming Attached: First relationships and how they shape our capacity to love.* Oxford University Press, 1994.

Kornfield, Jack. *The Wise Heart: A guide to the universal teachings of Buddhist psychology.* Bantam, 2009.

Lewis, Thomas, Fari Amini, and Richard Lannon. *A General Theory of Love.* Vintage Books, 2000.

MacDonald, Geoff, and Lauri A. Jensen-Campbell, editors. *Social Pain: Neuropsychological and health implications of loss and exclusion.* APA Press, 2011.

Mikulincer, Mario, and Phillip R. Shaver. *Attachment in Adulthood: Structure, dynamics, and change.* Guilford Press, 2007.

Putnam, Robert D. *Bowling Alone: The collapse and revival of American community.* Simon & Schuster, 2000.

Rifkin, Jeremy. *The Empathic Civilization.* Penguin, 2009.

Turkle, Sherry. *Alone Together: Why we expect more from technology and less from each other.* Basic Books, 2011.

Uchino, Bert. *Social Support and Physical Health: Understanding the health consequences of relationships.* Yale University Press, 2004.

Wright, Ronald. *A Short History of Progress.* House of Anansi Press, 2004.

PART ONE: The Relationship Revolution

Chapter 1: Love: A Paradigm Shift

Blum, Deborah. *Love at Goon Park: Harry Harlow and the science of affection.* Perseus Publishing, 2002.

Bowlby, John. *A Secure Base: Clinical applications of attachment theory*. Basic Books, 1988.

——. *Attachment and Loss, vol. 1: Attachment*. Basic Books, 1969.

——. *Attachment and Loss, vol. 2: Separation: Anxiety and anger*. Basic Books, 1973.

——. *Attachment and Loss, vol. 3: Sadness and depression*. Basic Books, 1981.

Buss, David, Todd Shackelford, Lee Kirkpatrick, and Randy Larsen. A half century of mate preferences: The cultural evolution of values. *Journal of Marriage and the Family,* 2001, vol. 63, pp. 491–503. Notes the changing criteria for choosing a mate.

Cacioppo, John, and William Patrick. *Loneliness: Human nature and the need for social connection*. Norton, 2008.

Chugani, Harry, Michael Behen, Otto Muzik, Csaba Juhasz, Ferenc Nagy, and Diane Chugani. Local brain functional activity following early deprivation: A study of post-institutionalized Romanian orphans. *Neuroimage,* 2001, vol. 14, pp. 1290–1301.

Darwin, Charles R. This is the question marry not marry [Memorandum on marriage—1838]. In *The Complete Work of Charles Darwin Online,* John Van Wyhe, editor, 2002. Available at http://darwin-online.org.uk/.

Descartes, René. *The Philosophical Writings of Descartes: Vol. 1*. Elizabeth Haldane and G. Ross, translators. Cambridge University Press, 1934.

Feeney, Brooke. The dependency paradox in close relationships: Accepting dependence promotes independence. *Journal of Personality and Social Psychology,* 2007, vol. 92, pp. 268–285. Notes the benefits of secure attachment for career women.

Fraley, Chris, David Fazzari, George Bonanno, and Sharon Dekel. Attachment and psychological adaptation in high exposure survivors of the September 11th attack on the World Trade Center. *Personality and Social Psychology Bulletin,* 2006, vol. 32, pp. 538–551.

Guenther, Lisa. *Social Death and Its Afterlives: A critical phenomenology of solitary confinement*. Minnesota University Press, 2012.

Hilgard, Ernest. *Psychology in America: A historical survey*. Harcourt Brace Jovanovich, 1993.

Johnson, Susan M. *Hold Me Tight: Seven conversations for a lifetime of love.* Little, Brown, 2008.

——. *The Practice of Emotionally Focused Couple Therapy: Creating connection.* Brunner/Routledge, 2004.

Karen, Robert. *Becoming Attached: First relationships and how they shape our capacity to love.* Oxford University Press, 1994. Karen notes data on "failure to thrive." For example, in 1915, 31 to 75 percent of children's deaths at ten eastern U.S. hospitals were attributed to failure to thrive.

McPherson, Miller, Lynn Smith-Lovin, and Matthew Brashears. Social isolation in America: Changes in core discussion networks over the two decades. *American Sociological Review,* 2006, vol. 71, pp. 353–375.

Mikulincer, Mario, and Phillip Shaver. *Attachment in Adulthood: Structure, dynamics, and change.* Guilford Press, 2007.

Sternberg, Robert. Triangulating love. In *The Psychology of Love,* Robert Sternberg and Michael Barnes, editors, pp. 119–138. Yale University Press, 1989.

Uchino, Bert. *Social Support and Physical Health: Understanding the health consequences of relationships.* Yale University Press, 2004.

Uchino, Bert, John Cacioppo, William B. Malarkey, Ronald M. Glaser, and Janice Kiecolt-Glaser. Appraised support predicts age-related differences in cardiovascular function in women. *Health Psychology,* 1995, vol. 14, pp. 556–562.

Whitehead, Barbara, and David Popenoe. *The State of Our Unions.* Report from Rutgers University National Marriage project, 2001. Notes that women reported that they wanted men who could share their feelings and that this was a priority for them. In 1967, by contrast, three-quarters of college women said they would marry a man they did not love if he met other criteria, mostly connected to the ability to support a family.

Chapter 2: Attachment: The Key to Love

Bartholomew, Kim, and Colleen Allison. An attachment perspective on abusive dynamics in intimate relationships. In *Dynamics of Romantic Love: Attachment, caregiving, and sex,* Mario Mikulincer and Gail

Goodman, editors, pp. 102–127. Guilford Press, 2006. Notes that anxiously attached male partners often become violent at breakup.

Blum, Deborah. *Love at Goon Park: Harry Harlow and the science of affection.* Perseus Publishing, 2002.

Bowlby, John. *A Secure Base: Clinical applications of attachment theory.* Basic Books, 1988.

——. *Attachment and Loss, vol. 1: Attachment.* Basic Books, 1969.

——. *Attachment and Loss, vol. 2: Separation: Anxiety and anger.* Basic Books, 1973.

——. *Forty-Four Juvenile Thieves: Their characters and home life.* Bailliere, Tindall & Cox, 1944.

Cassidy, Jude, and Phillip R. Shaver, editors. *Handbook of Attachment: Theory, research, and clinical implications.* Guilford Press, 2008.

Coan, James A., Hillary S. Schaefer, and Richard J. Davidson. Lending a hand: Social regulation of the neural response to threat. *Psychological Science,* 2006, vol. 17, pp. 1032–1039.

Davis, Deborah, Phillip R. Shaver, and Michael Vernon. Physical, emotional, and behavioral reactions to breaking up: The roles of gender, age, emotional involvement, and attachment style. *Personality and Social Psychology Bulletin,* 2003, vol. 29, pp. 871–884.

Hazan, Cindy, and Phillip R. Shaver. Love and work: An attachment-theoretical perspective. *Journal of Personality and Social Psychology,* 1990, vol. 59, pp. 270–280. Notes the three-category measure of attachment security used in the experiment at the end of the chapter.

——. Romantic love conceptualized as an attachment process. *Journal of Personality and Social Psychology,* 1987, vol. 52, pp. 511–524.

Johnson, Susan M. Attachment theory: A guide for healing couple relationships. In *Adult Attachment,* W. S. Rholes and J. A. Simpson, editors, pp. 367–387. Guilford Press, 2004.

Karen, Robert. *Becoming Attached: First relationships and how they shape our capacity to love.* Oxford University Press, 1998. Summarizes the history of attachment theory.

Lewis, Thomas, Fari Amini, and Richard Lannon. *A General Theory of Love.* Vintage Books, 2000.

Mikulincer, Mario, Gillad Hirschberger, Orit Nachmias, and Omri Gillath. The affective component of the secure base schema: Affective priming with representations of attachment security. *Journal of Personality and Social Psychology,* 2001, vol. 81, pp. 305–321.

Mikulincer, Mario, and Phillip R. Shaver. *Adult Attachment: Structure, dynamics, and change.* Guilford Press, 2007. Summarizes the basic concepts and supporting research of adult bonding.

Panksepp, Jaak. *Affective Neuroscience: The foundations of animal and human emotions.* Oxford University Press, 1998.

Sbarra, David. Predicting the onset of emotional recovery following non-marital relationship dissolution: Survival analyses of sadness and anger. *Personality and Social Psychological Bulletin,* 2006, vol. 32, pp. 298–312.

Simpson, Jeffry, Andrew Collins, SiSi Tran, and Katherine Haydon. Attachment and the experience and expression of emotions in romantic relationships: A developmental perspective. *Journal of Personality and Social Psychology,* 2007, vol. 92, pp. 355–367.

Simpson, Jeffry, William Rholes, and Julia Nelligan. Support seeking and support giving in couples in an anxiety provoking situation: The role of attachment styles. *Journal of Personality and Social Psychology,* 1992, vol. 62, pp. 434–446.

Simpson, Jeffry, William Rholes, and Dede Phillips. Conflict in close relationships: An attachment perspective. *Journal of Personality and Social Psychology,* 1996, vol. 71, pp. 899–914.

Suomi, Stephen. Attachment in rhesus monkeys. In *Handbook of Attachment,* Jude Cassidy and Phillip R. Shaver, editors, pp. 173–191. Guilford Press, 2008.

van den Boom, Dymphna. The influence of temperament and mothering on attachment and exploration: An experimental manipulation of sensitive responsiveness among lower-class mothers with irritable infants. *Child Development,* 1994, vol. 65, pp. 1457–1477.

Wilson, Edward. *Consilience: The unity of knowledge.* Vintage Books, 1998.

PART TWO: The New Science of Love

Chapter 3: The Emotions

Coan, James A., Hillary S. Schaefer, and Richard J. Davidson. Lending a hand: Social regulation of the neural response to threat. *Psychological Science,* 2006, vol. 17, pp. 1032–1039.

Damasio, Antonio. *Descartes' Error: Emotion, reason, and the human brain.* Putnam, 1994.

Darwin, Charles. *The Expression of Emotions in Man and Animals.* John Murray, 1872. Notes Darwin's habit of visiting the puff adder cage at the London zoo.

DeWall, C. Nathan, et al. Acetaminophen reduces social pain: Behavioral and neural evidence. *Psychological Science,* 2010, vol. 21, pp. 931–937.

Eisenberger, Naomi I., and Matthew D. Lieberman. Why rejection hurts: A common neural alarm system for physical and emotional pain. *Trends in Cognitive Science,* 2004, vol. 8, pp. 294–300.

Eisenberger, Naomi I., Matthew D. Lieberman, and Kipling D. Williams. Does rejection hurt? An fMRI study of social exclusion. *Science,* 2003, vol. 302, pp. 290–293.

Ekman, Paul. *Emotions Revealed.* Henry Holt, 2003.

Eugenides, Jeffrey. *Middlesex: A novel.* Farrar, Straus and Giroux, 2002.

Frederickson, Barbara L. The role of positive emotions in positive psychology: The broaden-and-build theory of positive emotions. *American Psychologist,* 2001, vol. 56, pp. 218–226.

Frijda, Nico H. *The Emotions.* Cambridge University Press, 1986.

Goldin, Philippe R., Kateri McRae, Wiveka Ramel, and James J. Gross. The neural bases of emotion regulation: Reappraisal and suppression of negative emotion. *Biological Psychiatry,* 2008, vol. 63, pp. 577–586.

Jack, Rachel E., Roberto Caldara, and Philippe G. Schyns. Internal representations reveal cultural diversity in expectations of facial expressions of emotion. *Journal of Experimental Psychology,* 2012, vol. 141, pp. 19–25. Notes evidence that culture affects the facial feature we focus on.

Johnson, Susan M., James A. Coan, M. M. Burgess, L. Beckes, A. Smith, T. Dalgleish, R. Halchuk, K. Hasselmo, P. S. Greenman, and Z. Merali. Soothing the threatened brain: Leveraging contact comfort with emotionally focused therapy. *PLOS One,* in press.

LeDoux, Joseph E. *The Emotional Brain: The mysterious underpinnings of emotional life.* Simon & Schuster, 1996.

Lehky, Sidney R. Fine discrimination of faces can be performed rapidly. *Journal of Cognitive Neuroscience,* 2000, vol. 12, pp. 848–855. Notes evidence that the detection and processing of the smallest change of expression on another's face occurs instantaneously.

Lieberman, Matthew, et al. Putting feelings into words: Affect labeling disrupts amygdala activity in response to affective stimuli. *Psychological Science,* 2007, vol. 18, pp. 421–428. Notes how the act of naming calms emotions.

Oatley, Keith. *Emotions: A brief history.* Blackwell, 2004.

Panksepp, Jaak. *Affective Neuroscience: The foundations of animal and human emotions.* Oxford University Press, 1998.

Schore, Alan N. *Affect Regulation and the Repair of the Self.* Norton, 2003.

Shaver, Phillip R., and Mario Mikulincer. Adult attachment strategies and the regulation of emotion. In *Handbook of Emotion Regulation,* James J. Gross, editor, pp. 446–465. Guilford Press, 2007.

Stenberg, Georg, Susanne Wiking, and Mats Dahl. Judging words at face value: Interference in a word processing task reveals automatic processing of affective facial expressions. *Cognition and Emotion,* 1998, vol. 12, pp. 755–782. Notes evidence that observers unconsciously mirror and synchronously (within 300 milliseconds) match the facial changes of another person.

Tronick, Edward Z. Dyadically expanded states of consciousness and the process of therapeutic change. *Infant Mental Health Journal,* 1998, vol. 19, pp. 290–299.

Young, Paul T. *Emotion in Man and Animal: Its nature and relation to attitude and motive.* Wiley, 1943. Notes the author's view that emotion involves the loss of cerebral control and all sense of conscious purpose.

Chapter 4: The Brain

Buchheim, Anna, et al. Oxytocin enhances the experience of attachment security. *Psychoneuroendocrinology,* 2009, vol. 34, pp. 1417–1422.

Campbell, Anne. Oxytocin and human social behavior. *Personality and Social Psychology Review,* 2010, vol. 14, pp. 281–295. Summarizes oxytocin's ability to enhance attachment and trust, improve social memory, and reduce fear.

Carter, C. Sue. Neuroendocrine perspectives on social attachment and love. *Psychoneuroendocrinology,* 1998, vol. 23, pp. 779–818. Summarizes the impact of oxytocin on bonding.

Carter, Rita, Susan Aldridge, Martyn Page, Steve Parker, and Chris Frith. *The Human Brain Book.* Dorling Kindersley, 2009. Offers an overview of the basic structure of the brain.

Chugani, Harry, Michael Behen, Otto Muzik, Csaba Juhasz, Ferenc Nagy, and Diane Chugani. Local brain functional activity following early deprivation: A study of post-institutionalized Romanian orphans. *Neuroimage,* 2001, vol. 14, pp. 1290–1301.

Cohen, Michael X., and Phillip R. Shaver. Avoidant attachment and hemispherical lateralization of the processing of attachment and emotion-related words. *Cognition and Emotion,* 2004, vol. 18, pp. 799–813. Notes that avoidant folks make more errors of interpretation, even in response to positive cues and emotions, than others do.

Cozolino, Louis J. *The Neuroscience of Human Relationships: Attachment and the developing social brain.* Norton, 2006.

de Waal, Frans. *The Age of Empathy: Nature's lessons for a kinder society.* Harmony Books, 2009.

Diamond, Lisa M. Contributions of psychophysiology to research on adult attachment: Review and recommendations. *Personality and Social Psychology Review,* 2001, vol. 5, pp. 276–295. Article summarizes the antistress effects of oxytocin.

Dick, Danielle M., et al. CHRM2, parental monitoring, and adolescent externalizing behavior: Evidence for gene environment interaction. *Psychological Science,* 2011, vol. 22, pp. 481–489.

Ditzen, Beate, Marcel Schaer, Barbara Gabriel, Guy Bodenmann, Ulrike

Ehlert, and Markus Heinrichs. Intranasal oxytocin increases positive communication and reduces cortisol levels during couple conflict. *Biological Psychiatry,* 2009, vol. 65, pp. 728–731.

Domes, Gregor, Markus Heinrichs, Andre Michel, Christoph Berger, and Sabine C. Herpertz. Oxytocin improves "mind-reading" in humans. *Biological Psychiatry,* 2007, vol. 61, pp. 731–733.

Ellis, Bruce J., Marilyn J. Essex, and W. Thomas Boyce. Biological sensitivity to context: II. Empirical explorations of an evolutionary-developmental theory. *Development & Psychopathology,* 2005, vol. 17, pp. 303–328. Notes the "orchid children" research.

Feeney, Brook, and Nancy Collins. Motivations for caregiving in adult intimate relationships: Influences on caregiving behavior and relationship functioning. *Personality and Social Psychology Bulletin,* 2003, vol. 29, pp. 950–968.

Freeman, Walter J. *How Brains Make Up Their Minds.* Weidenfeld & Nicolson, 1999.

Gillath, Omri, Silvia A. Bunge, Phillip R. Shaver, Carter Wendelken, and Mario Mikulincer. Attachment style differences in the ability to suppress negative thoughts: Exploring the neural correlates. *Neuroimage,* 2005, vol. 28, pp. 835–847.

Hebb, Donald. O. *The Organization of Behavior: A neuropsychological theory.* Wiley, 1949.

Hofer, Myron. Psychobiological roots of early attachment. *Current Directions in Psychological Science,* 2006, vol. 15, pp. 84–88. Notes that physiological regulation through the expression of emotion—for example, by a mother's soft voice and tender touch—shapes a child's general ways of regulating emotion. The child learns he can calm down and that emotion is workable.

Iacoboni, Marco. Imitation, empathy, and mirror neurons. *Annual Review of Psychology,* 2009, vol. 60, pp. 653–670.

———. *Mirroring People: The new science of how we connect with others.* Picador, 2008.

Joseph, Rhawn. Environmental influences on neural plasticity, the limbic system, emotional development, and attachment: A review. *Child*

Psychiatry and Human Development, 1999, vol. 29, pp. 189–208. Notes that social interactions grow our brains. Early institutionalization leads to decreased brain connectivity, especially between the amygdala and frontal cortex, the areas involved in development of social behavior in primates.

Kosfield, Michael, Markus Heinrichs, Paul J. Zak, Urs Fischbacher, and Ernst Fehr. Oxytocin increases trust in humans. *Nature,* 2005, vol. 435, pp. 673–676.

Liu, Dong, et al. Maternal care, hippocampal glucocorticoid receptors, and hypothalamic-pituitary-adrenal responses to stress. *Science,* 1997, vol. 277, pp. 1659–1661. Discusses research by Michael Meaney and his team.

Liu, Yan, Kimberly A. Young, J. Thomas Curtis, Brandon J. Aragona, and Zuoxin Wang. Social bonding decreases the rewarding properties of amphetamine through a dopamine D1 receptor-mediated mechanism. *The Journal of Neuroscience,* 2001, vol. 31, pp. 7960–7966.

Mikulincer, Mario, Tamar Dolev, and Phillip R. Shaver. Attachment-related strategies during thought suppression: Ironic rebounds and vulnerable self-representations. *Journal of Personality and Social Psychology,* 2004, vol. 87, pp. 940–956.

Mikulincer, Mario, and Phillip R. Shaver, editors. *Prosocial Motives, Emotions, and Behavior: The better angels of our nature.* APA Press, 2010. Notes the evolutionary function of empathy.

Pascual-Leone, Alvaro, and Roy Hamilton. The metamodal organization of the brain. In *Vision: From neurons to cognition,* vol. 134 of *Progress in Brain Research,* C. Casanova and M. Ptito, editors, pp. 427–445. Elsevier, 2001. Notes that the visual cortex processes auditory and tactile signals.

Powers, Sally I., Paula R. Pietromonaco, Meredith Gunlicks, and Aline Sayer. Dating couples' attachment styles and patterns of cortisol reactivity and recovery in response to a relationship conflict. *Journal of Personality and Social Psychology,* 2006, vol. 90, pp. 613–628. Notes that insecurity is linked to greater reactions to stress.

Rizzolatti, Giacomo, and Laila Craighero. The mirror-neuron system. *Annual Review of Neuroscience,* 2004, vol. 27, pp. 169–192.

Schore, Alan N. Effects of a secure attachment relationship on right brain development, affect regulation, and infant mental health. *Infant Mental Health Journal,* 2001, vol. 22, pp. 7–66.

Schwarz, Jaclyn M., Mark R. Hutchinson, and Staci D. Bilbo. Early-life experience decreases drug-induced reinstatement of morphine CPP in adulthood via microglial-specific epigenetic programming of anti-inflammatory IL-10 expression. *The Journal of Neuroscience,* 2011, vol. 31, pp. 17835–17847. Summarizes the Duke University study showing that touch impacts interleukin 10.

Singer, Tania. The neuronal basis and ontogeny of empathy and mind reading: Review of literature and implications for future research. *Neuroscience and Biobehavioral Reviews,* 2006, vol. 30, pp. 855–863.

Singer, Tania, Ben Seymour, John O'Doherty, Holger Kaube, Raymond Dolan, and Chris D. Frith. Empathy for pain involves the affective but not sensory components of pain. *Science,* 2004, vol. 303, pp. 1157–1162.

Stern, Daniel N. *The Present Moment in Psychotherapy and Everyday Life.* Norton, 2004.

Suomi, Stephen J. How gene-environment interactions shape biobehavioral development: Lessons from studies in rhesus monkeys. *Research in Human Development,* 2004, vol. 1, pp. 205–222.

———. Mother-infant attachment, peer relationships, and the development of social networks in rhesus monkeys. *Human Development,* 2005, vol. 48, pp. 67–79.

Theodoridou, Angelik, Angela C. Rowe, Ian S. Penton-Voak, and Peter J. Rogers. Oxytocin and social perception: Oxytocin increases perceived facial trustworthiness and attractiveness. *Hormones and Behavior,* vol. 56, pp. 128–132.

Tomasello, Michael, and Malinda Carpenter. Shared intentionality. *Developmental Science,* 2007, vol. 10, pp. 121–125. Notes that empathy helps cooperation.

315

Trevarthen, Colwyn. The functions of emotion in infancy: The regulation and communication of rhythm, sympathy, and meaning in human development. In *The Healing Power of Emotion,* Diana Fosha, Daniel J. Siegel, and Marion Solomon, editors, pp. 55–85. Norton, 2009. Discusses the concept of "prototypical conversations." The term was first used by the noted American anthropologist Mary Catherine Bateson.

Vicary, Amanda M., and R. Chris Fraley. Choose your own adventure: Attachment dynamics in a simulated relationship. *Personality and Social Psychology Bulletin,* 2007, vol. 33, pp. 1279–1291.

Chapter 5: The Body

ANSA (Agenzia Nazionale Stampa Associata). Italian men suffer "sexual anorexia" after Internet porn use. *La Gazzetta del Mezzogiorno.it,* March 4, 2013. Cites a recent survey conducted by Carlo Foresta of the Italian Society of Andrology and Sexual Medicine on Internet porn use and erectile dysfunction and loss of libido in men.

Basson, Rosemary. The female sexual response: A different model. *Journal of Sex and Marital Therapy,* 2000, vol. 26, pp. 51–65.

——. Women's sexual desire and arousal disorders. *Primary Psychiatry,* 2008, vol. 15, pp. 72–81.

Birnbaum, Gurit E. Attachment orientations, sexual functioning, and relationship satisfaction in a community sample of women. *Journal of Social and Personal Relationships,* 2007, vol. 24, pp. 21–35.

——. Beyond the borders of reality: Attachment orientations and sexual fantasies. *Personal Relationships,* 2007, vol. 14, pp. 321–342. Notes links between "solace sex" and sexual fantasies.

Birnbaum, Gurit E., Mario Mikulincer, and Omri Gillath. In and out of a daydream: Attachment orientations, daily couple interactions, and sexual fantasies. *Personality and Social Psychology Bulletin*, 2011, vol. 37, pp. 1398–1410. Summarizes the contribution of attachment orientations to the content of daily sexual fantasies.

Birnbaum, Gurit E., Harry T. Reis, Mario Mikulincer, Omri Gillath, and Ayala Orpaz. When sex is more than just sex: Attachment orientations, sexual experience, and relationship quality. *Journal of*

Personality and Social Psychology, 2006, vol. 91, pp. 929–943. Notes that attachment anxiety amplifies the effects of positive and negative sexual experiences on relationship interactions and that attachment avoidance inhibits the positive effects of having sex; also notes the detrimental effects of unsatisfying sex.

Bogaert, Anthony F., and Stan Sadava. Adult attachment and sexual behavior. *Personal Relationships,* 2002, vol. 9, pp. 191–204. Notes that secure partners feel more efficacious in sex: they believe the quality of the sex is mostly a reflection of them, not the situation.

Brennan, Kelly A., and Phillip R. Shaver. Dimensions of adult attachment, affect regulation, and romantic relationship functioning. *Personality and Social Psychology Bulletin,* 1995, vol. 21, pp. 267–283. Notes that those with a secure orientation prefer sex in a committed relationship and view expressing love as part of sex.

Brotto, Lori A. The DSM diagnostic criteria for hypoactive sexual desire disorder in women. *Archives of Sexual Behavior,* 2010, vol. 39, pp. 221–239. Notes that physical readiness for sex, such as engorgement of genital tissue, often does not translate into the experience of desire for women.

Castleman, Michael. Desire in women: Does it lead to sex? Or result from it? *Psychology Today* online, 2009. Available at http://www.psychol ogytoday.com/blog/all-about-sex/200907/desire-in-women-does-it-lead-sex-or-result-it.

Davis, Deborah, Phillip R. Shaver, and Michael L. Vernon. Attachment style and subjective motivations for sex. *Personality and Social Psychology Bulletin,* 2004, vol. 30, pp. 1076–1090.

Feeney, Judith A., Candida Peterson, Cynthia Gallois, and Deborah J. Terry. Attachment style as a predictor of sexual attitudes and behavior in late adolescence. *Psychology & Health,* 2000, vol. 14, pp. 1105–1122.

Gillath, Omri, and Melanie Canterbury. Neural correlates of exposure to subliminal and supraliminal sex cues. *Social Cognitive and Affective Neuroscience,* 2012, vol. 7, pp. 924–936.

Gillath, Omri, Mario Mikulincer, Gurit E. Birnbaum, and Phillip R.

Shaver. When sex primes love: Subliminal sexual priming motivates relationship goal pursuit. *Personality and Social Psychology Bulletin,* 2008, vol. 34, pp. 1057–1069.

Harding, Anne. Rosemary Basson: Working to normalize women's sexual reality. *The Lancet,* 2007, vol. 369, p. 363.

Hill, Craig A., and Leslie K. Preston. Individual differences in the experience of sexual motivation: Theory and measurement of dispositional sexual motives. *Journal of Sex Research,* 1996, vol. 33, pp. 27–45. Summarizes information on AMORE (Affective and Moral Orientation Related to Erotic Arousal), a scale for the measurement of motives for sex.

Hurlemann, René, et al. Oxytocin modulates social distance between males and females. *The Journal of Neuroscience,* 2012, vol. 32, pp. 16074–16079.

Impett, Emily A., Amie M. Gordon, and Amy Strachman. Attachment and daily sex goals: A study of dating couples. *Personal Relationships,* 2008, vol. 15, pp. 375–390. Notes that the very anxiously attached, even when they do not feel desire, are more likely to engage in sex because their partner wants it and that avoidant partners often engage in sex to avoid conflict and not to express love.

Insel, Thomas R., and Leslie E. Shapiro. Oxytocin receptor distribution reflects social organization in monogamous and polygamous voles. *Proceedings of the National Academy of Sciences,* 1992, vol. 89, pp. 5981–5985.

Johnson, Paul M., and Paul J. Kenny. Dopamine D2 receptors in addiction-like reward dysfunction and compulsive eating in obese rats. *Nature Neuroscience,* 2010, vol. 13, pp. 635–641.

Johnson, Susan M. *Hold Me Tight: Seven conversations for a lifetime of love.* Little, Brown, 2008. Summarizes, for the first time, synchrony, solace, and sealed-off sex as approaches to sex typified by different motivations and attachment orientations.

Kinsey, Alfred C., Wardell B. Pomeroy, and Clyde E. Martin. *Sexual Behavior in the Human Male.* W. B. Saunders Company, 1948.

Kinsey, Alfred C., Wardell B. Pomeroy, Clyde E. Martin, and Paul H.

Gebhard. *Sexual Behavior in the Human Female.* Indiana University Press, 1953.

Laan, Ellen, Walter Everaerd, Janneke van der Velde, and James H. Greer. Determinants of subjective experiences of sexual arousal in women: Feedback from genital arousal and erotic stimulus content. *Psychophysiology,* 1995, vol. 32, pp. 441–451.

Laumann, Edward O., John H. Gagnon, Robert T. Michael, and Stuart Michaels. *The Social Organization of Sexuality: Sexual practices in the United States.* University of Chicago Press, 1994. In the authors' national survey, 11 percent of women and 24 percent of men admitted engaging in extramarital sex.

Laumann, Edward O., and Robert T. Michael. *Sex, Love, and Health in America: Private choices and public policies.* University of Chicago Press, 2001.

Laumann, Edward O., Anthony Paik, and Raymond C. Rosen. Sexual dysfunction in the United States: Prevalence and predictors. *Journal of the American Medical Association,* 1999, vol. 281, pp. 537–544.

Maltz, Wendy, and Larry Maltz. *The Porn Trap: The essential guide to overcoming problems caused by pornography.* HarperCollins, 2008.

Marcus, I. David. Men who are not in control of their sexual behavior. In *Handbook of Clinical Sexuality for Mental Health Professionals,* S. Levine, C. Risen, and S. Althof, editors, pp. 383–400. Brunner/Routledge, 2003. Notes that 6 to 8 percent of men are addicted to sex.

Masters, William H., and Virginia E. Johnson. *Human Sexual Response.* Bantam, 1966. *Human Sexual Inadequacy.* Little, Brown, 1970. These books summarize the first laboratory studies of the anatomy and physiology of human sexual response and define the four stages of sex as desire, arousal, orgasm, and resolution.

McCarthy, Barry, and Emily McCarthy. *Rekindling Desire.* Brunner/Routledge, 2003.

McCarthy, Barry W., and Michael E. Metz. *Men's Sexual Health: Fitness for satisfying sex.* Routledge, 2007.

Meston, Cindy M., and David M. Buss. Why humans have sex. *Archives of Sexual Behaviour,* 2007, vol. 36, pp. 477–507.

Meyers, Laurie. The Eternal Question: Does love last? APA *Monitor,* 2007, vol. 38, pp. 44–47. Contains quotation by Elaine Hatfield.

Michael, Robert T., John H. Gagnon, Edward O. Laumann, and Gina Kolata. *Sex in America: A definitive survey.* Little, Brown, 1994.

Tiefer, Leonore. Sexual behaviour and its medicalisation. *British Medical Journal,* 2002, vol. 325, p. 45. Notes that sex can be viewed as digestion or as a dance between intimates.

TopTenReviews. Internet Pornography Statistics, 2013. Available at: http://internet-filter-review.toptenreviews.com/internet-pornography-statistics.html.

PART THREE: Love in Action

Chapter 6: Love across Time

AARP (formerly American Association of Retired Persons). *The Divorce Experience: A study of divorce at midlife and beyond.* Conducted for *AARP The Magazine.* AARP, 2004.

Atkinson, Leslie, Susan Goldberg, Vaishali Ravel, David Pederson, Diane Benoit, Greg Moran, Lori Poulton, Natalie Myhal, Michael Zwiers, Karin Gleason, and Eman Leung. On the relation between maternal state of mind and sensitivity in the prediction of infant attachment security. *Developmental Psychology,* 2005, vol. 41, pp. 42–53.

Beckes, Lane, Jeffry A. Simpson, and Alyssa Erickson. Of snakes and succor: Learning secure attachment associations with novel faces via negative stimulus pairings. *Psychological Science,* 2010, vol. 21, pp. 721–728.

Brown, Susan, and Lin I-Fen. *Divorce in Middle and Later Life: New estimates from the 2008 American community survey.* Bowling Green State University. Notes the prevalence of divorce in mature adults.

Cowan, Carolyn Pape, and Philip A. Cowan, *When Partners Become Parents: The big life change for couples.* Erlbaum Associates, 2000.

Eastwick, Paul W., and Eli J. Finkel. The attachment system in fledgling relationships: An activating role for attachment anxiety. *Journal of Personality and Social Psychology,* 2008, vol. 95, pp. 628–647.

References

Friedman, Howard S., and Leslie R. Martin. *The Longevity Project: Surprising discoveries for health and long life from the landmark eight-decade study.* Penguin, 2011.

Gottman, John M., and Julie Schwartz Gottman. *And Baby Makes Three: The six-step plan for preserving marital intimacy and rekindling romance after baby arrives.* Crown, 2007. Notes that marital satisfaction drops in two-thirds of couples when the first child arrives.

Hall, Scott S., and Rebecca A. Adams. Cognitive coping strategies of newlyweds adjusting to marriage. *Marriage & Family Review,* 2011, vol. 47, pp. 311–325.

Huston, Ted L., John P. Caughlin, Renate M. Houts, Shanna E. Smith, and Laura J. George. The connubial crucible: Newlywed years as predictors of marital delight, distress, and divorce. *Journal of Personality and Social Psychology,* 2001, vol. 80, pp. 237–252.

Mancini, Anthony D., and George A. Bonanno. Marital closeness, functional disability, and adjustment in late life. *Psychology and Aging,* 2006, vol. 21, pp. 600–610.

McLean, Linda M., Tara Walton, Gary Rodin, Mary J. Esplen, and Jennifer M. Jones. A couple-based intervention for patients and caregivers facing end-stage cancer: Outcomes of a randomized controlled trial. *Psycho-Oncology,* 2013, vol. 22, pp. 28–38.

Misri, Shaila, Xanthoula Kostaras, Don Fox, and Demetra Kostaras. The impact of partner support in the treatment of postpartum depression. *The Canadian Journal of Psychiatry,* 2000, vol. 45, pp. 554–558.

Ravel, Vaishali, Susan Goldberg, Leslie Atkinson, Diane Benoit, Natalie Myhal, Lori Poulton, and Michael Zwiers. Maternal attachment, infant responsiveness, and infant attachment. *Infant Behavior and Development,* 2001, vol. 24, pp. 281–304. Notes that anxious and avoidant attachment styles in mothers affect how they care for their children and how secure their children become.

Rholes, W. Steven, Jamie L. Kohn, A. McLeish Martin III, Jeffry A. Simpson, Carol L. Wilson, SiSi Tran, and Deborah A. Kashy. Attachment orientations and depression: A longitudinal study of new parents. *Journal of Personality and Social Psychology,* 2011, vol. 100, pp. 567–586.

Rosand, Gun-Mette B., Kari Slinning, Malin Eberhard-Gran, Espen Roysamb, and Kristian Tamb. Partner relationship satisfaction and maternal emotional distress in early pregnancy. *BMC Public Health,* 2011, vol. 11, pp. 161–173. Notes the impact of postpartum depression.

Shapiro, Alyson F., and John M. Gottman. Effects on marriage of a psycho-communication-educational intervention with couples undergoing transition to parenthood: Evaluation at one year post-intervention. *Journal of Family Communication,* 2005, vol. 5, pp. 1–24.

Shapiro, Alyson F., John M. Gottman, and Sybil Carrere. The baby and the marriage: Identifying factors that buffer against decline in marital satisfaction after the first baby arrives. *Journal of Family Psychology,* 2000, vol. 14, pp. 59–70.

Taylor, Paul, et al. *The Decline of Marriage and Rise of New Families.* Pew Research Center, 2010.

Uchino, Bert N., John T. Cacioppo, and Janice Kiecolt-Glaser. The relationship between social support and physiological processes: A review with emphasis on underlying mechanisms and implications for health. *Psychological Bulletin,* 1996, vol. 119, pp. 488–531.

Chapter 7: Unraveling Bonds

Bowlby, John. *The Making and Breaking of Affectional Bonds.* Routledge, 1979. Notes the concept of deprivation.

Eastwick, Paul W., and Eli J. Finkel. Sex differences in mate preferences revisited: Do people really know what they initially desire in a romantic partner? *Journal of Personality and Social Psychology,* 2008, vol. 94, pp. 245–264.

Eldridge, Katherine A., and Andrew Christensen. Demand-withdraw communication during couple conflict: A review and analysis. In *Understanding Marriage,* Patricia Noller and Judith Feeney, editors, pp. 289–322. Cambridge University Press, 2002.

Feeney, Judith. Hurt feelings in couple relationships: Towards integrative models of negative effects of hurtful events. *Journal of Social and Personal Relationships,* 2004, vol. 21, pp. 487–508.

Fine, Cordelia. *Delusions of Gender: How our minds, society, and neurosexism create difference.* Norton, 2010.

Finkel, Eli J., Paul W. Eastwick, Benjamin Karney, Harry Reis, and Susan Sprecher. Online dating: A critical analysis from the perspective of psychological science. *Psychological Science,* 2012, vol. 13, pp. 3–66.

Gottman, John. *Seven Principles for Making Marriage Work.* Crown, 1999. Notes that stonewalling predicts divorce.

Gudrais, Elizabeth. When words hurt: How depression lingers. *Harvard Magazine,* July–August 2009. Notes the quote from Jill Hooley.

Heavey, Christopher L., Christopher Layne, and Andrew Christensen. Gender and conflict structure in marital interaction II: A replication and extension. *Journal of Consulting and Clinical Psychology,* 1993, vol. 61, pp. 16–27.

Herman, Judith Lewis. *Trauma and Recovery.* Basic Books, 1992, p. 54. Notes that abuse by attachment figures is a "violation of human connection."

Hooley, Jill M., and Ian H. Gotlib. A diathesis-stress conceptualization of expressed emotion and clinical outcome. *Applied and Preventative Psychology,* 2000, vol. 9, pp. 135–151.

Hooley, Jill M., Staci A. Gruber, Laurie A. Scott, Jordan B. Hiller, and Deborah A. Yurgelun-Todd. Activation in dorsolateral prefrontal cortex in response to maternal criticism and praise in recovered depressed and healthy control participants. *Biological Psychiatry,* 2005, vol. 57, pp. 809–812.

Hooley, Jill M., and John D. Teasdale. Predictors of relapse in unipolar depressives: Expressed emotion, marital distress, and perceived criticism. *Journal of Abnormal Psychology,* 1989, vol. 98, pp. 229–235.

Huston, Ted L., John P. Caughlin, Renate M. Houts, Shanna E. Smith, and Laura J. George. The connubial crucible: Newlywed years as predictors of marital delight, distress, and divorce. *Journal of Personality and Social Psychology,* 2001, vol. 80, pp. 237–252.

Hyde, Janet Shibley. The gender similarities hypothesis. *American Psychologist,* 2005, vol. 60, pp. 581–592.

Ickes, William. *Everyday Mind Reading: Understanding what other people think and feel.* Prometheus Books, 2003.

Johnson, Susan M. *Emotionally Focused Therapy with Trauma Survivors: Strengthening attachment bonds*. Guilford, 2002.

Johnson, Susan M., Judy A. Makinen, and John W. Millikin. Attachment injuries in couple relationships: A new perspective on impasses in couple therapy. *Journal of Marital and Family Therapy,* 2001, vol. 27, pp. 145–155.

Pasch, Lauri A., and Thomas N. Bradbury. Social support, conflict, and the development of marital dysfunction. *Journal of Consulting and Clinical Psychology,* 1998, vol. 66, pp. 219–230.

Roberts, Linda J., and Danielle R. Greenberg. Observational windows to intimacy processes in marriage. In *Understanding Marriage: Developments in the study of couple interaction,* Patricia Noller and Judith Feeney, editors, pp. 118–149. Cambridge University Press, 2002.

Tronick, Ed. *The Neurobehavioral and Social Emotional Development of Infants and Children*. Norton, 2007.

Chapter 8: Renewing Bonds

Burgess-Moser, Melissa, Susan M. Johnson, Tracy L. Dalgleish, and Giorgio Tasca. The impact of blamer softening on romantic attachment in emotionally focused therapy. *Journal of Marital and Family Therapy.* Study in review.

Cohen, Shiri, Robert J. Waldinger, Marc S. Schulz, and Emily Weiss. Eye of the beholder: The individual and dyadic contributions of empathic accuracy and perceived empathic effort to relationship satisfaction. *Journal of Family Psychology,* 2012, vol. 26, pp. 236–245.

Cohn, Jeffrey F., Reinaldo Matias, Edward Z. Tronick, David Connell, and Karlen Lyons-Ruth. Face-to-face interactions of depressed mothers and their infants. *New Directions for Child Development,* 1986, vol. 34, pp. 31–45.

Diamond, Lisa M. Contributions of psychophysiology to research on adult attachment: Review and recommendations. *Personality and Social Psychology Review,* 2001, vol. 5, pp. 276–295.

Greenman, Paul S., and Susan M. Johnson. Process research on Emotionally Focused Therapy (EFT) for couples: Linking theory to practice. *Family Process,* 2013, vol. 52, pp. 46–61.

Halchuk, Rebecca E., Judy A. Makinen, and Susan M. Johnson. Resolving attachment injuries in couples using emotionally focused therapy: A three-year follow-up. *Journal of Couple and Relationship Therapy,* 2010, vol. 9, pp. 31–47.

Johnson, Susan M. *Hold Me Tight: Conversations for connection. A facilitator's guide.* Ottawa Couple and Family Institute, 2009.

Makinen, Judy A., and Susan M. Johnson. Resolving attachment injuries in couples using EFT: Steps towards forgiveness and reconciliation. *Journal of Consulting and Clinical Psychology,* 2006, vol. 74, pp. 1055–1064.

Murray, Sandra L., Dale W. Griffin, Jaye L. Derrick, Brianna Harris, Maya Aloni, and Sadie Leder. Tempting fate or inviting happiness? Unrealistic idealization prevents the decline of marital satisfaction. *Psychological Science,* 2011, vol. 22, pp. 619–626.

Roberts, Linda J., and Danielle R. Greenberg. Observational windows to intimacy processes in marriage. In *Understanding Marriage: Developments in the study of couple interaction,* Patricia Noller and Judith Feeney, editors, pp. 118–149. Cambridge University Press, 2002.

Salvatore, Jessica E., Sally I-Chun Kuo, Ryan D. Steele, Jeffry A. Simpson, and W. Andrew Collins. Recovering from conflict in romantic relationships: A developmental perspective. *Psychological Science,* 2011, vol. 22, pp. 376–383.

Tronick, Edward Z., and Jeffrey F. Cohn. Infant-mother face-to-face interaction: age and gender differences in coordination and the occurrence of miscoordination. *Child Development,* 1989, vol. 60, pp. 85–92.

Zuccarini, Dino, Susan M. Johnson, Tracy L. Dalgleish, and Judy A. Makinen. Forgiveness and reconciliation in emotionally focused therapy for couples: The client change process and therapist interventions. *Journal of Marital and Family Therapy,* 2013, vol. 39, pp. 148–162. Available at http://onlinelibrary.wiley.com/doi/10.1111/j.1752-0606.2012.00287.x/abstract.

PART FOUR: The New Science Applied

Chapter 9: A Love Story

Burgess-Moser, Melissa, Susan M. Johnson, Tracy L. Dalgleish, and Giorgio Tasca. The impact of blamer softening on romantic attachment in emotionally focused therapy. *Journal of Marital and Family Therapy.* Study in review.

Greenman, Paul S., and Susan M. Johnson. Process research on EFT for couples: Linking theory to practice. *Family Process,* 2013, vol. 52, pp. 46–61.

International Center of Excellence in Emotionally Focused Therapy. EFT Research, 2006, online at http://www.iceeft.com/EFTResearch.pdf. Contains a summary of EFT outcome and process-of-change research over the past twenty-five years.

Johnson, Susan M. Attachment theory: A guide for healing couple relationships. In *Adult Attachment,* W. S. Rholes and J. A. Simpson, editors, pp. 367–387. Guilford Press, 2004.

———. *Hold Me Tight: Seven conversations for a lifetime of love.* Little, Brown, 2008.

———. *The Practice of Emotionally Focused Couple Therapy: Creating connection.* Brunner/Routledge, 2004.

Johnson, Susan M., and Paul Greenman. The path to a secure bond. *Journal of Clinical Psychology: In Session,* 2006, vol. 62, pp. 597–609.

Lebow, Jay L., Anthony L. Chambers, Andrew Christensen, and Susan M. Johnson. Research on the treatment of couple distress. *Journal of Marital and Family Therapy,* 2012, vol. 38, pp. 145–168.

Makinen, Judy A., and Susan M. Johnson. Resolving attachment injuries in couples using EFT: Steps towards forgiveness and reconciliation. *Journal of Consulting and Clinical Psychology,* 2006, vol. 74, pp. 1055–1064.

Chapter 10: Love in the 21st Century

Administration for Children and Families (ACF). *The Healthy Marriage Initiative: Myths and facts.* US Department of Health and Human Services, 2005.

References

Broidy, Lisa M., Richard E. Tremblay, et al. Developmental trajectories of childhood disruptive behaviors and adolescent delinquency: A six-site, cross-national study. *Developmental Psychology,* 2003, vol. 39, pp. 222–245.

Cacioppo, John, and William Patrick. *Loneliness: Human nature and the need for social connection.* Norton, 2008.

Caldwell, Benjamin E., Scott R. Woolley, and Casey J. Caldwell. Preliminary estimates of cost-effectiveness of marital therapy. *Journal of Marital & Family Therapy,* 2007, vol. 33, pp. 392–405. Notes the costs of divorce.

Caprara, Gian V., Claudio Barbaranelli, Concetta Pastorelli, Albert Bandura, and Philip G. Zimbardo. Prosocial foundations of children's academic achievement. *Psychological Science,* 2000, vol. 11, pp. 302–306.

Certain, Laura. Report on Television and Toddlers. Presented at the Pediatric Academic Societies annual meeting, Baltimore, April 30, 2003.

Davidovitz, Rivka, Mario Mikulincer, Phillip R. Shaver, Ronit Izsak, and Micha Popper. Leaders as attachment figures: Leaders' attachment orientations predict leadership-related mental representations and followers' performance and mental health. *Journal of Personality and Social Psychology,* 2007, vol. 93, pp. 632–650.

Dex, Shirley, and Colin Smith. *The Nature and Pattern of Family-Friendly Employment Policies in Britain.* Judge Institute of Management, Cambridge University: The Policy Press, 2002.

Diener, Ed, Richard Lucas, Ulrich Schimmack, and John F. Helliwell. *Well-Being for Public Policy.* Oxford University Press, 2009.

Dion, M. Robin. Healthy marriage programs: Learning what works. *The Future of Children,* 2005, vol. 15, pp. 139–156. Notes and describes the main marriage education programs used in the United States.

Earle, Alison, Zitha Mokomane, and Jody Heymann. International perspectives on work-family policies: Lessons from the world's most competitive economies. *The Future of Children,* 2011, vol. 21, pp. 191–210. Notes that allowing people to take family-related leave actually helps economies and does not raise expenditures.

Flores, Philip J. *Addiction as an Attachment Disorder*. Jason Aronson/Rowman & Littlefield, 2004.

Gordon, Mary. *Roots of Empathy: Changing the world child by child*. Thomas Allen, 2007.

Grunfeld, Eva, Robert Glossop, Ian McDowell, and Catherine Danbrook. Caring for elderly people at home: The consequences to caregivers. *Canadian Medical Association,* 1997, vol. 157, pp. 1101–1105.

Halford, W. Kim. *Marriage and Relationship Education: What works and how to provide it*. Guilford Press, 2011.

Hamilton, Brady E., Joyce A. Martin, and Stephanie J. Ventura. Births: Preliminary data for 2007. *National Vital Statistics Reports,* vol. 57. [US] National Center for Health Statistics, 2009.

Harter, James K., Frank L. Schmidt, and Corey L. M. Keyes. Well-being in the workplace and its relationship to business outcomes: A review of the Gallup studies. In *Flourishing: The positive person and the good life,* Corey Keyes and Jonathon Haidt, editors, pp. 205–224. American Psychological Association, 2003.

Jacobs, Jane. *The Death and Life of Great American Cities*. Random House, 1961.

Kasser, Tim, Richard M. Ryan, Melvin Zax, and Arnold J. Sameroff. The relations of maternal and social environments to late adolescents' materialistic and prosocial values. *Developmental Psychology,* 1995, vol. 31, pp. 907–914.

Kingston, Anne, with Alex Ballingall. Public display of disaffection: As "cell-fishness" hits an all-time high, a backlash against mobile devices includes outright bans. *Maclean's,* September 15, 2011. Available at www2.macleans.ca/2011/09/15/public-display-of-disaffection/.

Klinenberg, Eric. I want to be alone: The rise and rise of solo living. *The Guardian,* March 30, 2012.

Kraus, Michael W., Cassie Huang, and Dacher Keltner. Tactile communication, cooperation, and performance: An ethological study of the NBA. *Emotion,* 2010, vol. 10, pp. 745–749.

Levy, Daniel. *Love and Sex with Robots: The evolution of human-robot relationships*. HarperCollins, 2007.

References

Logsdon, Mimia C., and Angela B. McBride. Social support and post-partum depression. *Research in Nursing & Health,* 1994, vol. 17, pp. 449–457.

McGregor, Holly A., Joel D. Leiberman, Jeff Greenberg, Sheldon Solomon, Jamie Arndt, Linda Simon, and Tom Pyszczynski. Terror management and aggression. *Journal of Personality and Social Psychology,* 1998, vol. 74, pp. 590–605.

Mikulincer, Mario, and Phillip R. Shaver. Attachment theory and inter-group bias: Evidence that priming secure base scheme attenuates negative reactions to out-groups. *Journal of Personality and Social Psychology,* 2001, vol. 81, pp. 97–115.

———. Boosting attachment security to promote mental health, prosocial values, and inter-group tolerance. *Psychological Inquiry,* 2007, vol. 18, pp. 139–156.

Mikulincer, Mario, Phillip R. Shaver, Omri Gillath, and Rachel A. Nitzberg. Attachment, caregiving, and altruism: Boosting attachment security increases compassion and helping. *Journal of Personality and Social Psychology,* 2005, vol. 89, pp. 817–839.

Myers, Scott M., and Alan Booth. Marital strains and marital quality: The role of high and low locus of control. *Journal of Marriage and Family,* 1999, vol. 61, pp. 423–436.

Neal, David T., and Tanya L. Chartand. Embodied emotion perception: Amplifying and dampening facial feedback modulates emotion perception accuracy. *Social Psychological and Personality Science,* 2011, vol. 2, pp. 673–678.

Olweus, Dan. *Bullying at School: What we know and what we can do about it.* Blackwell, 1993.

Ooms, Theodora, Stacey Bouchet, and Mary Parke. *Beyond Marriage Licenses: Efforts in states to strengthen marriage and two-parent families.* Center for Law and Social Policy, 2004.

Putnam, Robert D. *Bowling Alone: The collapse and revival of American community.* Simon & Schuster, 2000.

Romano, Elisa, Richard E. Tremblay, Frank Vitaro, Mark Zoccolillo, and Linda Pagani. Prevalence of psychiatric diagnosis and the role of perceived impairment: Findings from an adolescent community sample. *Journal of Child Psychology and Psychiatry,* 2001, vol. 42, pp. 451–461. Notes that one in five young people experience mental health problems.

Schramm, David G. Individual and social costs of divorce in Utah. *Journal of Family and Economic Issues,* 2006, vol. 27, pp. 133–151.

Sennett, Richard. *Together: The rituals, pleasures, and politics of cooperation.* Yale University Press, 2012.

Turkle, Sherry. *Alone Together: Why we expect more from technology and less from each other.* Basic Books, 2011.

UNICEF. Child poverty in perspective: An overview of child well-being in rich countries. *Innocenti Report Card 7,* 2007. Innocenti Research Centre, Florence, Italy.

Wright, Ronald. *A Short History of Progress.* House of Anansi Press, 2004.

Index

ABOUT THE AUTHOR

Dr. Sue Johnson is director of the International Centre for Excellence in Emotionally Focused Therapy, distinguished research professor in couple and family therapy at Alliant International University in San Diego, California, and professor of clinical psychology at the University of Ottawa, Canada. She is a fellow of the American Psychological Association and has received numerous honors for her work, including the Outstanding Contribution to Marriage and Family Therapy award from the American Association for Marriage and Family Therapy and the Distinguished Contribution to Family Therapy Research award from the American Family Therapy Academy. She is a frequent guest speaker at professional meetings around the globe and trains mental health professionals worldwide in adult attachment and bonding and couple and family interventions. She also is a consultant to numerous organizations, including the U.S. and Canadian military, the U.S. Department of Veterans Affairs, and the New York City Fire Department. She loves medieval history, Argentine tango, birding, kayaking, and traveling to exotic locales.